Designing RF Combining Systems for Shared Radio Sites

For a listing of recent titles in the
Artech House Mobile Communications Library,
turn to the back of this book.

Designing RF Combining Systems for Shared Radio Sites

Ian Graham

ARTECH HOUSE
BOSTON | LONDON
artechhouse.com

Library of Congress Cataloging-in-Publication Data
A catalog record for this book is available from the U.S. Library of Congress

British Library Cataloguing in Publication Data
A catalog record for this book is available from the British Library.

ISBN-13: 978-1-68569-025-0

Cover design by Joi Garron

685 Canton St.
Norwood, MA

10 9 8 7 6 5 4 3 2 1

To the memory of Ron Keefe, principal engineer, and Bill Mitchell, engineering manager; without your belief and investment in me during my time as a young, raw RF engineer in the Airadio Lab at Marconi Avionics in the early 1980s, I would never have had the career and opportunities that I have enjoyed since; and to the memory of Angus Tait, whose sense of community and determination to succeed on the world stage continue to inspire all who work at Tait Communications

Contents

Preface

In this book, the reader will learn how to design interference-free repeater site radio frequency (RF) systems that maximize reliable coverage in mission-critical land mobile radio (LMR) networks.

The book was written to address a gap in the available literature, as there are very few books or industry standards on this subject. The relevance is shown as the time of writing coincided with the U.S. Federal Communications Commission (FCC) issuing Policy Statement 23-122 concerning the need to address growing levels of interference in radio systems. One recommendation highlighted is that "radio systems should use good engineering practice to mitigate degradation from interference," meaning that the FCC is going to start looking at your radios and your system design if you have an interference problem, and your system will need to be up to scratch if you wish to be protected. In that respect, designing your repeater site RF systems using the method provided in this book will place you in a good position.

Now let us bring it closer to home; imagine you are part of the team designing a six-site VHF P25 Phase 1 LMR network that will provide all the police, fire, and ambulance mission-critical communications for a small city and its surrounding areas. Your task is to design the RF antenna and combining systems at the six repeater sites in such a way that they deliver maximum reliable coverage. All six of the repeater sites are known to have other radio systems already operating there, some of which occupy the same part of the very high frequency (VHF) band that you will be using. To deliver the coverage that the emergency services require, you will have to ensure that the antenna and combining systems that you are designing do not cause or suffer from interference as a result of these colocated systems. What do you do? Where do you even begin? By the end of this book, you will know exactly how to approach the situation and solve the problem.

The book covers the complete repeater site RF antenna and combining system design process in order, noting that successful LMR system design is increasingly dependent on early identification and mitigation of the potential interference mechanisms that could occur.

The Introduction describes what LMR networks are and the type of organizations that typically use them. The basic architecture of an LMR network and its operation are described. Then the concept of interference and its growing effect on the ability of the users to communicate reliably is introduced. The reasons why interference is most likely to occur at the repeater sites are outlined, showing why that is the area of the system on which we concentrate in this book.

Chapter 1 then gives an overview of the repeater site RF systems themselves. The basic architecture consisting of an antenna system with an associated transmitter combiner and receiver multicoupler is introduced and the function of each of these parts described.

The worst thing you can do when designing a new repeater site RF system for deployment on an existing site is to only consider the system you are designing. Instead, before starting any design, detailed site surveys need to be carried out so the engineer can understand what other RF systems are already installed at the site and the RF environment that results. How to do this is covered in Chapter 2.

Chapter 3 provides technical background on noise and intermodulation in radio systems so that the reader will understand how they cause interference in radio communications systems later in the book.

Chapter 4 expands on the noise and intermodulation background in Chapter 3 to describe the types of interference that they cause at repeater sites. These are the interference mechanisms that we will be mitigating by design in later chapters.

The remainder of the book covers the repeater site RF antenna and combining system design process in order, starting with Chapter 5, which describes the types of antennas used at repeater sites and then goes on to show how to design an optimal antenna system for your system. Design rules for best practice antenna system design are given at the end of the chapter.

Having discussed the types of interference that can occur at repeater sites, Chapter 6 shows how to calculate the level to which these must be suppressed in our repeater site RF system so that they do not cause noticeable receiver deafening and loss of uplink (talk-in) coverage.

In Chapter 7, the reader will learn how to design the simplest type of repeater site RF system where a single transmitter and receiver are connected to the antenna via a duplexer. We will learn what duplexers are and how to specify the filtering required to prevent interference.

Chapters 8 and 9 deal with the design of more complicated multichannel repeater site RF systems. Chapter 8 focuses on transmitter combiners that allow multiple base station transmitters to be coupled onto a single transmit antenna. We will learn how to minimize losses, maximize radiated RF power, and specify the filtering performance required to prevent any of the base station receivers colocated at the site from being deafened.

Chapter 9 concentrates on receiver multicouplers that enable the associated base station receivers to be fed from one receive antenna. The reader will learn about the various types of receiver multicouplers, when to use them, how to specify the filtering required to prevent interference, and how to set up the receiving system for optimum uplink coverage. In a similar fashion to Chapter 5, both Chapters 8 and 9 end by providing best practice design rules.

Repeater sites are often at geographical high points and so are exposed to extreme conditions. This makes the failure of the equipment installed outdoors, like the antennas, a very real possibility during the lifetime of the system. Chapter 10 describes a couple of ways of remotely monitoring the health of your antenna systems so that impending failure can be identified and addressed before it happens.

That covers the design process in order, but once your repeater site RF system is designed, you need to connect it all together by means of coaxial cables. How to do this is the subject of Chapter 11 and it is not as trivial as it sounds because choosing the right cable for each part of the system can have a positive impact on system performance and durability. Grounding and how to protect your system from lightning strikes are also discussed in this chapter.

Chapter 12 is the last and longest chapter in the book, intended to bring together everything in the book in a step-by-step worked example. This chapter is the complete design solution for one of the sites in the emergency services P25 Phase 1 system example described earlier. We will look at how to use the information gathered in the site and RF surveys to design the site-specific, interference-free antenna and combining systems required.

Introduction

After spending a career in the highly rewarding field of designing reliable land mobile radio communication systems that can literally save lives, my hope in this book is to share some of that experience with the readership. Although this book is written with an LMR bias, the physics is the same across all branches of radio communications, meaning that the techniques and design methodologies described are just as applicable to high-performance military radio networks, cellular Long-Term Evolution (LTE) systems, and even Wi-Fi networks.

LMR networks are terrestrially based, professional, push-to-talk wireless systems that provide critical communications for public safety agencies such as police, fire, and other first responders. Such networks are increasingly used in the commercial world too, by essential infrastructure organizations including transportation, electricity, oil, and gas.

Figure I.1 shows a high-level block diagram of a typical LMR network.

The touch screen dispatch and web-based IP dispatch console are used by an operator who works for an organization such as the police or fire department to broadcast messages over the LMR network and send members of the organization to the places where they are needed. Smartphones and IP phones can also be used to do this if they have a push-to-talk (PTT) application that enables them to behave like any other radio connected to the network. The function of the LMR gateway is to enable all of these to be connected to the radio network. Many public safety LMR networks are designed to meet the

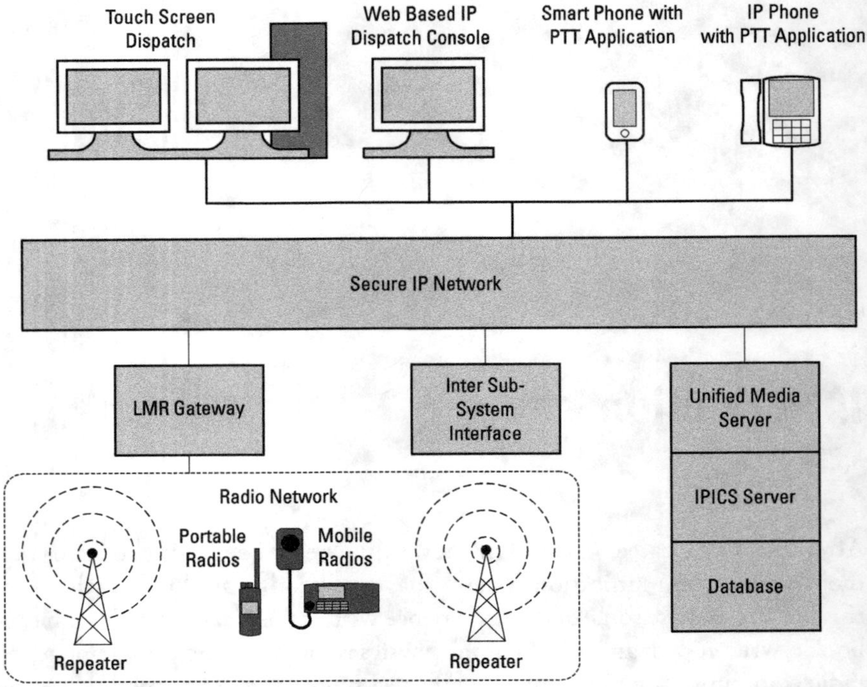

Figure I.1 Typical LMR network block diagram.

Association of Public-Safety Communications Officials (APCO) P25 standard, so it is the role of the Inter Subsystem Interface to allow one P25 system to connect to another (quite likely designed by a different vendor) wherever necessary. The Unified Media Server and the IP Interoperability and Collaboration System (IPICS) Server facilitate the interconnection between the radio network and the dispatcher consoles and phones. Finally, the radio network itself consists of handheld portable radios and vehicle-mounted mobile radios through which the users communicate via a number of suitably located fixed base station repeater systems located at repeater sites spaced throughout the area to be covered.

Large-scale basic LMR analog voice systems have been deployed since the 1930s. Since the mid-1980s, public safety agencies and critical infrastructure organizations have progressively migrated to more advanced digital LMR systems that provide a greater call capacity and support both voice and data communications. Additionally, many of these users have upgraded from basic conventional systems to more complex trunked systems. Conventional

LMR systems have dedicated frequencies and channels assigned to individual groups of users, so when a user manually selects a specific channel and makes a call, the other members of the group cannot use that channel until the call is over. Therefore, in a conventional LMR network, the number of users that can be supported is directly proportional to the number of channels available. Experience dictates that this is usually fine for organizations with up to three groups of fewer than 70 users. Trunked systems are more complex but allow for more efficient sharing of channels among larger groups of users. A trunked system has a pool of channels, and when a call is made by a user, an available channel from this pool is automatically assigned by the network. Although trunked LMR networks are more complicated than conventional systems, the advantages of increased call capacity and reduced congestion can more than cover the additional cost. Trunked systems take advantage of the fact that not everyone will need a channel at the same time, so the number of users that can be supported grows exponentially with the number of channels, as shown in Figure I.2.

For many public safety or critical infrastructure organizations, their LMR networks provide the only secure communications during critical situations, often in remote areas well beyond the reach of commercial cellular networks. The radio network can be the only means of communication that workers in potentially dangerous situations have back to the outside world; it can literally be their lifeline and that is why it is so important that we, as RF system engineers, get the design right.

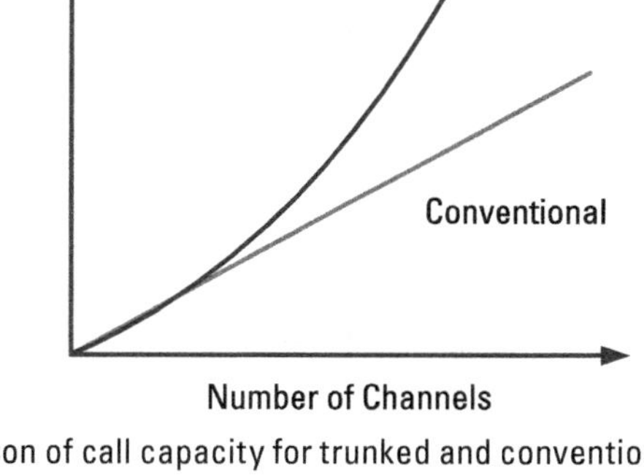

Figure I.2 Comparison of call capacity for trunked and conventional LMR networks.

When an organization like this buys an LMR system, what they are effectively buying is coverage, the ability to communicate reliably over their entire service area. The area of the radio spectrum used by LMR networks (136–941 MHz) is becoming ever more congested as the number of radio systems crowded into that space increases. So, an increasingly demanding, yet fundamental, aspect in the design of a reliable LMR network is interference mitigation, essentially designing your LMR network so it is not deafened by the many other radio systems operating in the same area. Interference is the enemy of coverage; when a radio system is affected by interference, sudden and unexpected coverage loss is the result. Sometimes, this is not even apparent to the user relying on the system, so they will think their important message has been heard when it has not.

Obviously, the more radio systems that you try to operate at similar frequencies within a given geographic area, the more potential for interference there is. To gain some insight, if we look more closely at the radio network elements from Figure I.1, we will see in that this has three main parts:

- Radio products: the base stations and mobile and portable radios;
- A coverage design that accurately determines the number of sites necessary to cover the required service area, which could be anything from a town or city, to an entire country;
- Repeater systems: RF systems at fixed radio sites involving the base stations, antennas, and combining systems; these are what actually provide the coverage for the users to communicate.

These are shown in Figure I.3.

The radio network consists of handheld portable radios and vehicle-mounted mobile radios through which the users communicate via the fixed base station repeater systems located at sites appropriately spaced throughout the service area to be covered.

Looking at these in turn, the better the RF performance of the base stations and mobile and portable radios is, the greater the distance they will be able to communicate over, meaning that we need fewer sites to cover the service area. Also, the communication will be more reliable than would be the case if cheaper radios with inferior RF performance were used. The use of more expensive, better performing base station and mobile and portable radio products from a reputable vendor will often result in a more economical network once the cost of the sites and their repeater systems is considered too.

Next is the coverage design, usually in the form of a computer simulation that predicts how far acceptable communication can be achieved from

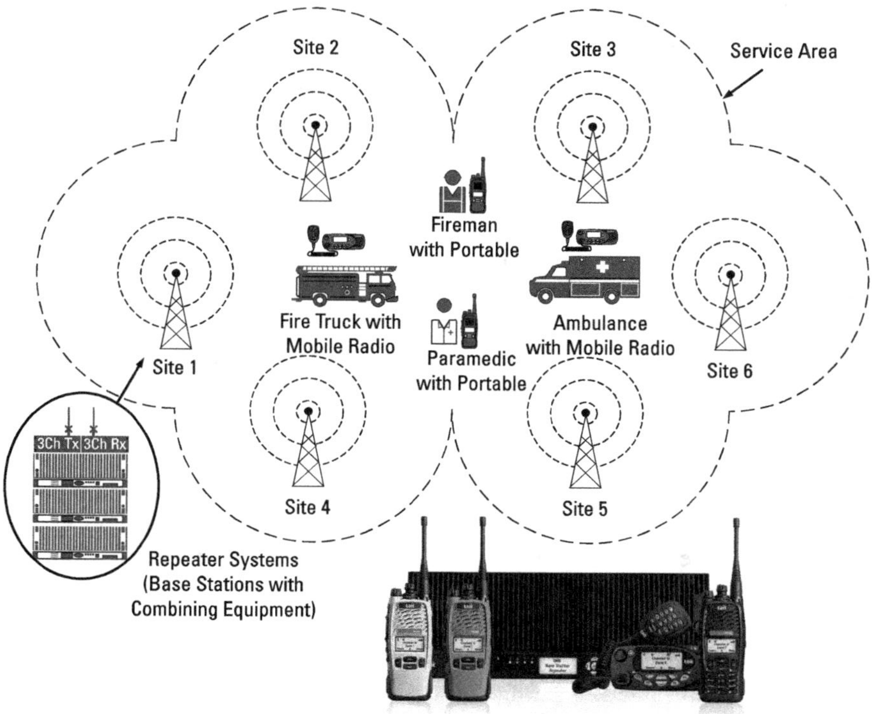

Figure I.3 The radio network part of an LMR system.

each of the repeater sites. The use of realistic propagation models, antenna radiation patterns, and terrain/clutter maps are fundamental aspects of being able to accurately determine the number of repeater sites necessary to meet the contractual coverage goals in terms of guaranteeing a minimum acceptable quality of service with a stated reliability (usually 90% or 95%) over the entire service area.

Finally, there are the RF systems on the repeater sites themselves. These consist of a number of base stations (however many are required to serve the number of users in that part of the service area), and combining equipment required to plumb the base station transmitters and receivers up to typically one transmit antenna and one receive antenna on the radio tower, all of which have to perform optimally in the presence of the many other radio systems already operating at the site.

Designing the radio part of an LMR network is the art of choosing the right radio products to use and designing the repeater site RF systems so that when the complete network is built and tested, it performs as predicted in the coverage design. In other words, the radio part of the LMR network provides the performance required to satisfy the end users communication requirements.

This book concentrates on the design of the repeater site RF systems because the sites are the place where the sensitive LMR receivers are exposed to the greatest number of high-level interfering signals from nearby transmitters. In other words, in an LMR network, the sites are the place where interference is most likely to occur and result in unexpected coverage loss. Repeater sites are usually located at geographic high points because you get the greatest coverage, so most organizations operating radio systems in the area will want to install their repeaters there. However, the point is that no matter how well the base stations and mobile and portable radios perform, or how accurate the coverage design was, the performance of the entire system can be crippled by poor repeater site RF system design.

For example, interference and unexpected coverage loss are very likely results if you think only of the new system that you want to deploy on a radio site without due attention to the transmit and receive frequencies of all the other radio systems already installed there. The same goes for where you place your antennas on the radio towers; this must be done with respect to where the existing antennas are. An LMR system designed in isolation is unlikely to meet the contractual coverage and reliability goals, and these shortcomings will not be revealed until very late in the project cycle, during coverage verification testing, site acceptance testing, or, worse still, when the system is being actively relied upon by the end user. Expensive and time-consuming fault finding, redesign, and contractual penalties are the probable consequences. The necessary investigations take a frustratingly long time and usually result in many unbudgeted dollars having to be spent to correct matters.

Additionally, once interference at radio sites starts to occur, it just gets progressively worse as further radio systems are deployed at the site, and eventually things can reach the point where the whole site becomes unusable.

So, the ability to design repeater site RF systems that do not cause or suffer from interference, but instead fit in seamlessly with the other radio systems already in use at the site is a fundamental requirement of any robust LMR network. The trick is to treat each repeater site as a system in its own right. Therefore, before starting any work on any new repeater site RF system design, information must be gathered about that site so that the RF system design engineer can understand the RF environment present. That is, understand exactly what other RF systems are already installed at the site, what the transmit and receive frequencies of each of them are, and where their respective antennas are on the tower (or towers). Only once the RF system engineers have this information can they carry out informed frequency planning, decide where to locate their antennas on the towers, and design combining systems that ensure that the repeater site's RF system performs optimally.

A project that I worked on in the mid-2010s is a good example of the problems that interference at radio repeater sites can cause in an LMR network. The customer, a U.S. sheriff's department, operated an LMR network used by all the fire, police, ambulance, and search-and-rescue agencies in a large city and the surrounding areas. For several years, they had had problems reliably making calls, and each time they addressed this by adding more channels to their repeater site systems at the 20 sites that covered the service area. However, what happened each time they did this was that the reliability problems got worse, not better. They eventually opted to refresh all their LMR radio infrastructure, and the company for which I was working was selected as the vendor. Not long into the project, once RF surveys had been carried out on the radio sites, it became obvious that their problem was not capacity (the number of channels installed at each site), but interference due to other radio systems that shared the sites. Once this was understood and their repeater site RF systems were redesigned properly to mitigate these interference problems, reliable communications were reestablished over the entire service area and with fewer channels.

In a nutshell, that is what this book is about: how to design repeater site RF systems that do not cause or suffer from interference because of the other radio systems already installed at the site and thereby deliver the goal of maximum reliable coverage.

Finally, while explaining things purely in terms of mathematics has its place, this book is written empirically with the explanation first and the necessary calculation afterwards.

1

Repeater Site RF Systems Overview

In land mobile radio (LMR) networks, the repeater site radio frequency (RF) systems basically receive transmissions from users in the field and retransmit (repeat) them to the intended recipient. Most LMR systems will have a network of repeater sites, appropriately located to provide acceptable coverage over the desired area. It all sounds simple enough, and it is, relatively, until you consider that each of these favorably located repeater sites is likely to have a number of other independent radio systems also installed there. As described in the Introduction, that is where things start to get a little more complicated once these independent radio systems start interfering with each other.

Let us keep things simple for now though and look at the various types of repeater site RF systems and how they work. There are basically two main types.

1.1 Single-Channel Duplexed System

First, if only one radio channel is needed to provide service for the users in the area covered by that site, then the duplexed repeater site RF system shown in Figure 1.1 is all that is required.

This is the simplest form of repeater site RF system, consisting only of one base station, an isolator, and a duplexer connected to a common transmit and receive antenna. The duplexer is a passive device consisting of two filters in

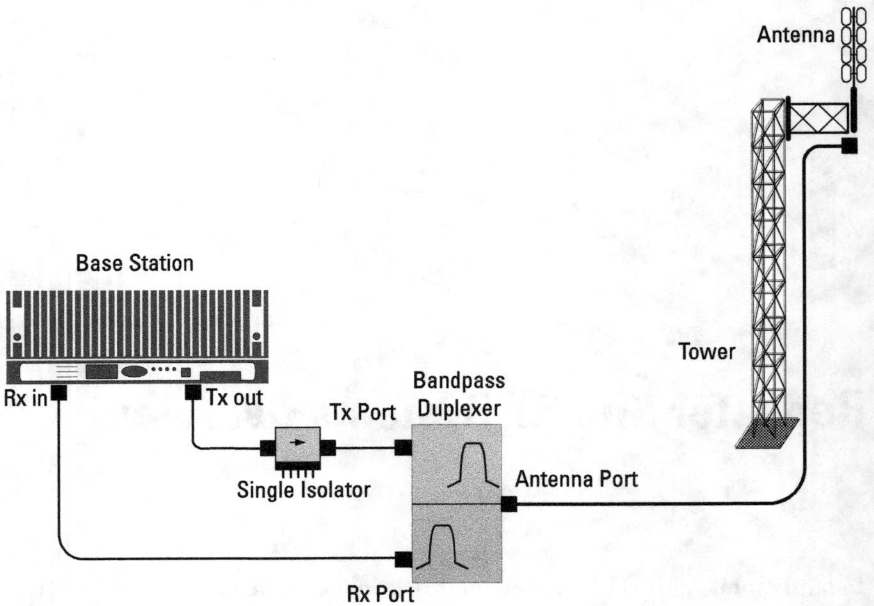

Figure 1.1 Single-channel duplexed repeater site RF system.

parallel. One filter passes the transmitted signal to the antenna for transmission, and the other passes the desired signal to the receiver and rejects many of the unwanted signals also picked up by the antenna. The passbands of the transmit and receive filters do not overlap, thereby isolating the base station's transmitter output from its receiver input. The various types of duplexers and when or when not to use each one will be covered in more detail in Chapter 7. The only other part in the single-channel system is the isolator. This is a ferrite device that only allows RF to pass through in one direction, with any reflected energy being terminated in the isolator's dump load. This protects the transmitter output stages where reflected energy from a fault further towards the antenna could otherwise cause damage and subsequent failure.

1.2 Multichannel System

If several radio channels are required on a site to provide service to a greater number of users, then the required repeater site RF system is more complicated, as shown in Figure 1.2. This multichannel configuration is the most common form of repeater site RF system because modern LMR networks are increasingly required to offer high call capacity. Within a multichannel site RF system, it is not feasible for each transmitter and receiver to have its own dedicated antenna; we would quickly run out of space on the tower. Instead, the normal compromise is to have one transmit antenna and one receive antenna. The receive antenna is mounted uppermost to maximize the distance over which users in the field can communicate with the site. The transmit antenna is mounted a few tens of feet directly below to achieve the required isolation between them (how to calculate the required transmit to receive antenna separation is covered in Chapter 5).

With such an antenna system, we need a cost-effective method allowing us to combine multiple base station transmitters onto the one transmit antenna that:

- Provides identical coverage for all the connected transmitters;
- Keeps the loss of transmit power to a minimum;
- Cleans up all the transmitted signals enough to prevent interference with any receiver on the site;
- Can be easily expanded with further transmitters if the call capacity needs increasing in future.

A transmitter combining system that provides these functions usually consists of a number of bandpass filters, one for each transmitter, as shown in Figure 1.2. Each filter is tuned to the transmit frequency of the specific base station to which it is connected, and the outputs are then connected together with coaxial cables. Isolators are again connected in each base station transmitter output to protect the transmitters in the event of a fault further towards the antenna. In this case, though, dual isolators consisting of two isolators cascaded together are used to also minimize the generation (and subsequent radiation) of intermodulation products that would otherwise occur if each transmitter signal fed into the outputs of neighboring transmitters on the combiner. Later in the book, Chapter 8 describes the various types of transmitter combiners and how to design them.

Similarly, we also need a method that allows the one receive antenna to feed multiple base station receivers. We need a way of doing this that:

- Provides identical coverage no matter which base station receiver is used;
- Prevents any loss of performance (and therefore coverage) as a result of dividing the received signal between multiple receivers;
- Prevents any transmitter on the site from deafening any of the connected receivers;

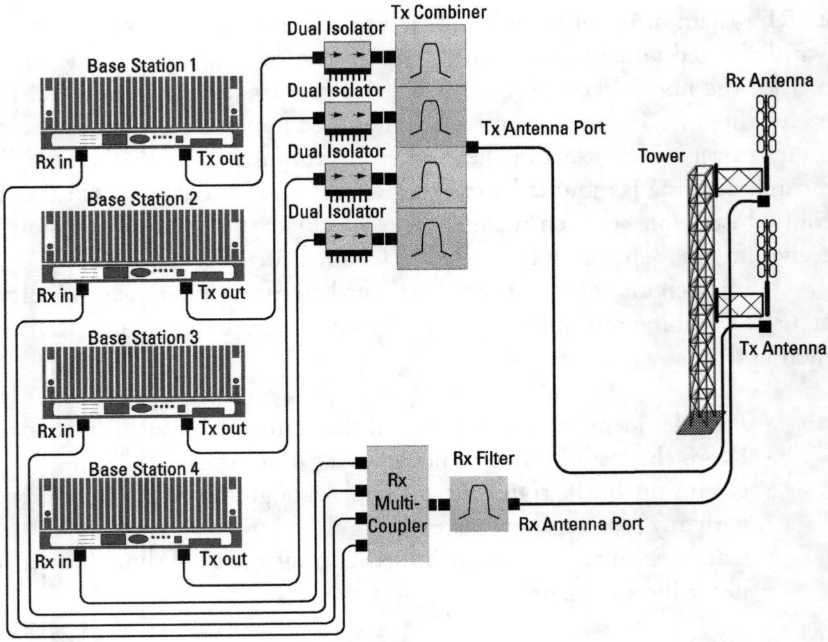

Figure 1.2 Multichannel repeater site RF system.

- Can be easily expanded with further receivers if the call capacity needs increasing in future.

The area of the system shown in Figure 1.2 that addresses these needs is the receiver multicoupler. The first part of this is the receiver filter, or preselector, that passes the weak desired signals but protects the individual base station receivers from the strong interfering signals from nearby site transmitters. This filter is followed by a low noise amplifier to counteract the electrical losses and noise figures of the other stages in the receiving system. Next is a variable attenuator that allows the overall gain to be set correctly, and, finally, there is a splitter that divides the signal equally between the connected receivers. Chapter 9 describes the various types of receiver multicouplers, how to design them, and how to set them up depending on the RF environment at the site.

Finally, try to avoid designing multichannel duplexed systems. Sometimes, we are forced to because there is not space on the tower for two antennas or because of tower loading concerns. However, avoid this type of system if possible, as it makes interference management much more difficult, as will be explained in Chapter 7.

2

Surveying the Sites

Designing an RF system for a repeater site is not just a matter of considering the new design that you want to deploy there. Instead, you need to treat the site as a system in its own right and do your design with respect to the radio systems already operating in the locality. So, before doing any design work, it is necessary to gather the information that enables you to fully understand both the construction readiness and the local RF environment at each of the sites in your LMR system.

Record-keeping of what radio equipment is installed at any given repeater site and where the respective antennas are located is regulated to an extent. The licenses required to operate LMR transmitters also record valuable information about frequency, radiated power level, and emission bandwidth. However, as the part of the radio spectrum occupied by the LMR system gets more congested, the information legally required to be kept often falls short of what we need when carrying out a new LMR system design. In any one LMR system design, you will tend to find that excellent records are kept at some repeater sites that you are using, while only the bare legal minimum is written down at others.

It follows that the quality of the information you have available prior to starting the design of your repeater site RF systems will be directly proportional to the performance of the resulting LMR system. Providing the required overall coverage relies on all the sites performing optimally, the consequence of basing some of your repeater site RF system designs on incomplete information will be interference that unacceptably reduces the overall system coverage.

It really does take significantly less time to find the information required to do the design properly than it does to fault find and correct the suboptimal design that will likely result otherwise. So, in this chapter, we will explain exactly what information is needed before starting any new repeater site RF system design work and how to go about collecting it.

The best way to collect the necessary information is to visit the sites and carry out the four surveys listed here:

- Site survey;
- Tower mapping;
- RF survey;
- Tower structural survey.

The surveys are divided up in this way because each one is likely to be carried out by a different person or organization, each with specific expertise. Even at sites where excellent records are kept, there is no substitute for going there and seeing it for yourself; only by doing that will you really get a sense of how complete and up to date the record-keeping is.

2.1 Site Survey

The site survey is a physical inspection conducted at each site with the purpose of cataloging the existing RF assets and evaluating the construction readiness. There are recommendations but no industry standards detailing what information should be obtained. So, it is important to define what you expect from the survey, as it may be either someone from your company, the customer's organization, or an independent third-party specialist that actually carries it out. As a bare minimum, the site survey should provide the following information:

- Latitude and longitude coordinates of the site;
- Elevation above sea level;
- Site plans showing locations, dimensions, and orientation of equipment shelters, towers, ice bridges, fences, power and telephone poles, and any generators or transformers;
- Dimensioned equipment shelter plans showing entry and exit points, where each equipment rack is located within the shelter, and the associated cable trays;

- Layout drawings of each rack in the equipment shelter showing the name of the system, the base stations (including Tx and Rx frequencies), transmitter combiners, receiver multicouplers, and any additional filters or duplexers;
- Details of all AC/DC power systems on the site and the circuit breaker locations;
- Details of the grounding systems inside and outside the shelter;
- Photographs of the site from all sides and all faces of the tower;
- Details and photographs of near-end obstructions that may affect the propagation of RF signals, including the surrounding tree line, nearby tall buildings, and areas of higher terrain.

2.2 Tower Mapping

It is here that we get into more detail about each antenna on the tower of interest, its function, and the transmitters or receivers that they map back to in the equipment shelter. As a minimum, tower mapping should deliver the following information:

- The latitude and longitude coordinates of the tower;
- The height of the tower, orientation relative to true north, and the size of each face;
- Photographs of each section of each face of the tower;
- A drawing of each face of the tower showing and numbering each antenna;
- For each numbered antenna, list:
  - The leg or face of the tower on which the antenna is mounted;
  - The size and type (e.g., colinear, exposed dipole, Yagi, panel, microwave dish);
  - The model number and frequency range;
  - The mounting height, above ground level;
  - The azimuth relative to true north;
  - The lengths of any standoffs used to separate the antenna from the tower leg;
  - The attached transmission line type and diameter;
  - The marking, color coding, or labeling of each transmission line;
  - The port of entry into or out of the equipment shelter.
- Mapping of receive antennas via the filter and multicoupling equipment to the respective base station receivers in the equipment shelter;

- Mapping of transmit antennas via the combining equipment to the respective base station transmitters in the equipment shelter;
- Mapping of duplexed antennas via the duplexer to the respective base station transmitter and receiver in the equipment shelter.

Note that the final two points enable us to search for the regulatory licenses pertaining to each site transmitter and thereby find the RF power radiated from the associated antenna.

If there is only one tower on site and no other towers in the vicinity, the information gathered from the site survey and tower mapping will be enough information to enable us to estimate the frequency and level of each interfering signal that will be present at the Rx antenna as shown in Figure 2.1. From that, we can then determine the most sensible locations to place our new transmit and receive antennas on the tower, as will be described in the design rules in Chapter 5.

2.3 RF Surveys

Commonly though, there is more than one tower present on conveniently located geographic high sites. On sites like this, rather than attempting to map each tower and then work out all the interfering signal levels, it is easier to carry out an RF survey and physically measure the interfering signals that your proposed receive antenna will be exposed to as shown in Figure 2.2. (Note that you still need to map the tower onto which you are deploying your antennas to decide the best available locations.)

Again, there is no actual standard that defines the duration of an RF survey. Generally, the longer, the better, but this will be limited by how many sites there are in your LMR system and how long you have to do the design. A sensible duration for an RF survey is 24 hours during a weekday, to ensure that you get the busy hour. The RF survey itself is usually divided into two parts: (1) spectrum monitoring and (2) noise monitoring.

2.3.1 Spectrum Monitoring

Spectrum monitoring is where we define and monitor the band of interest for the recommended duration to pick up any existing RF activity that may affect our proposed new repeater site RF system design. The band monitored will usually be the LMR band within which the new system will reside, plus an extra 10% above and below so that you pick up all the unwanted signals

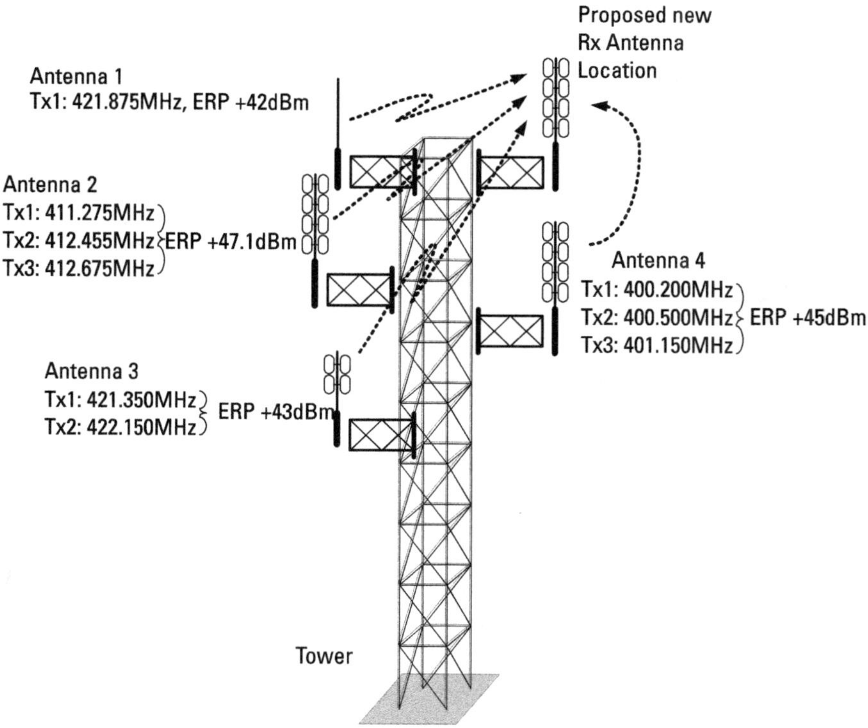

Antenna	Transmitter Frequency (MHz)	Effective Radiated Power (dBm)	Isolation to RX Antenna (dB)	Interferer Level at Rx Antenna (dBm)
1	421.875	+42	25	+17
2	411.275	+47.1	45	+2.1
	412.455			
	412.675			
3	421.350	+43	62	–19
	422.150			
4	400.200	+45	64	–19
	400.500			
	401.150			

Figure 2.1 Tower mapping to determine interfering signal frequencies and levels.

that could cause interference. For example, if the new RF system that we were designing will operate in the 150–155-MHz band, then it makes sense to monitor the site for existing RF activity from 135 to 170 MHz.

A suitable test setup for spectrum monitoring is shown in Figure 2.3. It uses an antenna that is sensitive across the whole band that you want to

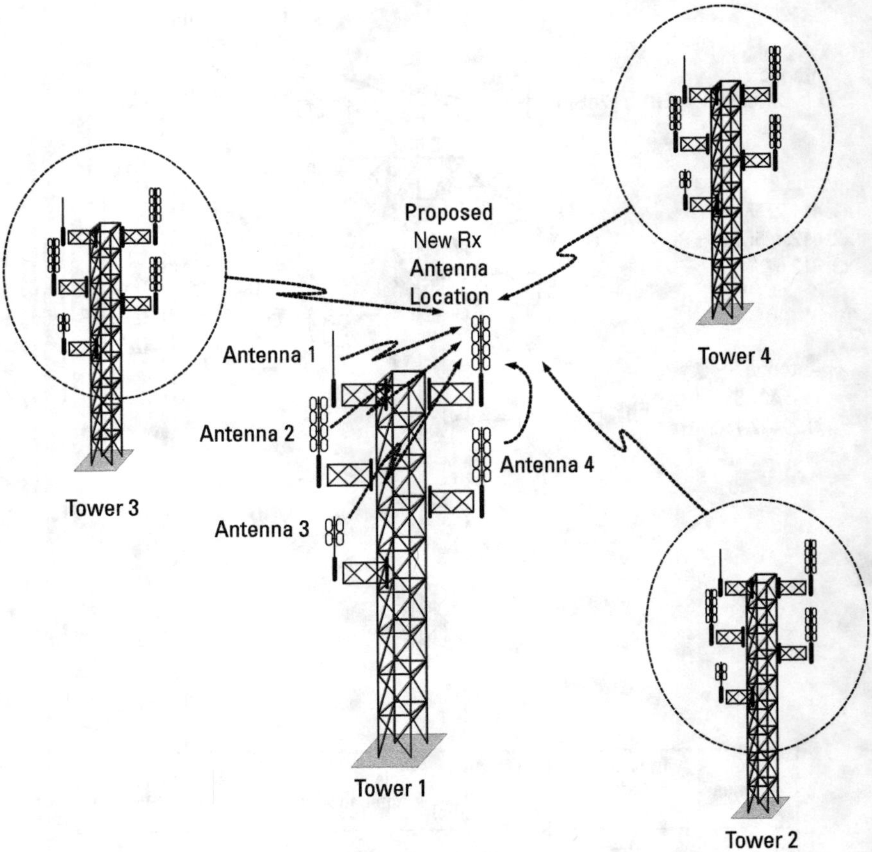

Figure 2.2 Interference from colocated towers at the site.

monitor, placed as close as possible to the tower where your proposed new receive antenna will be located. Placing the monitoring antenna here will mean that you will get an accurate picture of the interference your system will be exposed to from all the other towers at the site. The antenna is then connected to a bandpass filter that passes signals within the monitored band, while suppressing any high-level out-of-band signals (e.g., broadcast signals that could overload the monitoring receiver). The output of the bandpass filter is connected to a spectrum analyzer that measures the frequency and level of the existing RF signals present within the monitored band. The spectrum analyzer is then interfaced to a laptop computer that controls the test and logs the frequency and level of the signals picked up and then calculates their occupancy (percentage of time present).

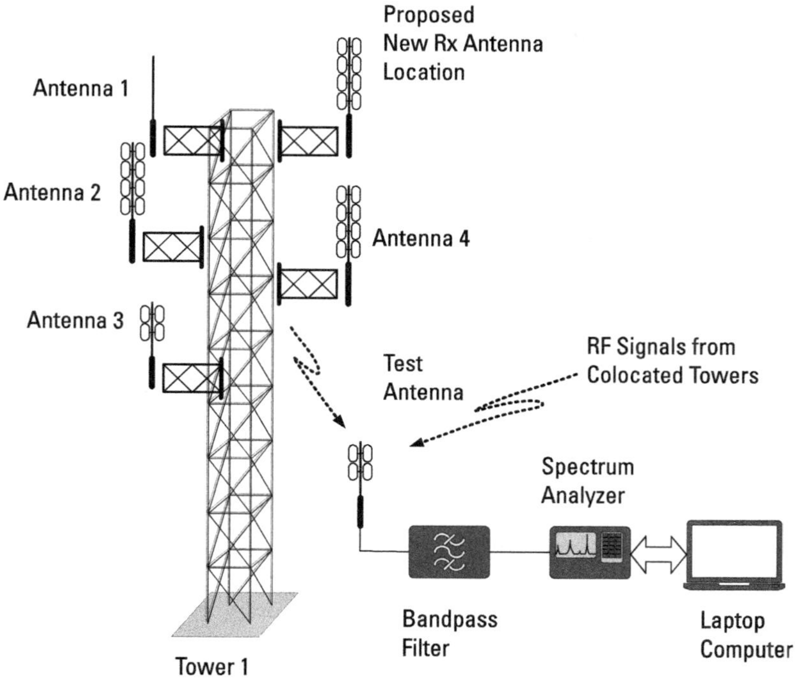

Figure 2.3 Spectrum monitoring test setup.

Once the test is completed, a spectrum report needs to be issued detailing all the RF activity detected within the band of interest over the period monitored. For each RF signal picked up within the band of interest, the frequency (to within +/–1 kHz), the level, and the occupancy as a percentage of time against a threshold of –80 dBm need to be recorded. Figure 2.4 gives a graphical example of this.

From the spectrum monitoring results, we can determine which of the signals picked up have the potential to create harmful interference. My experience of mitigating interference in the repeater site RF systems has shown that existing signals picked up at a level of –40 dBm for greater than 5% of time are the ones that have the potential to contribute towards intermodulation (IM) that will adversely impact the LMR system performance. Such signals are termed high-level carriers (HLCs). Except for unwanted signals in one of the same channels that you plan to use (cochannel interference) or in a channel next to one that you plan to use (adjacent channel interference), any other unwanted signals that are below –40 dBm in level or present less than 5% of the time will have little or no impact on system operation. Table 2.1

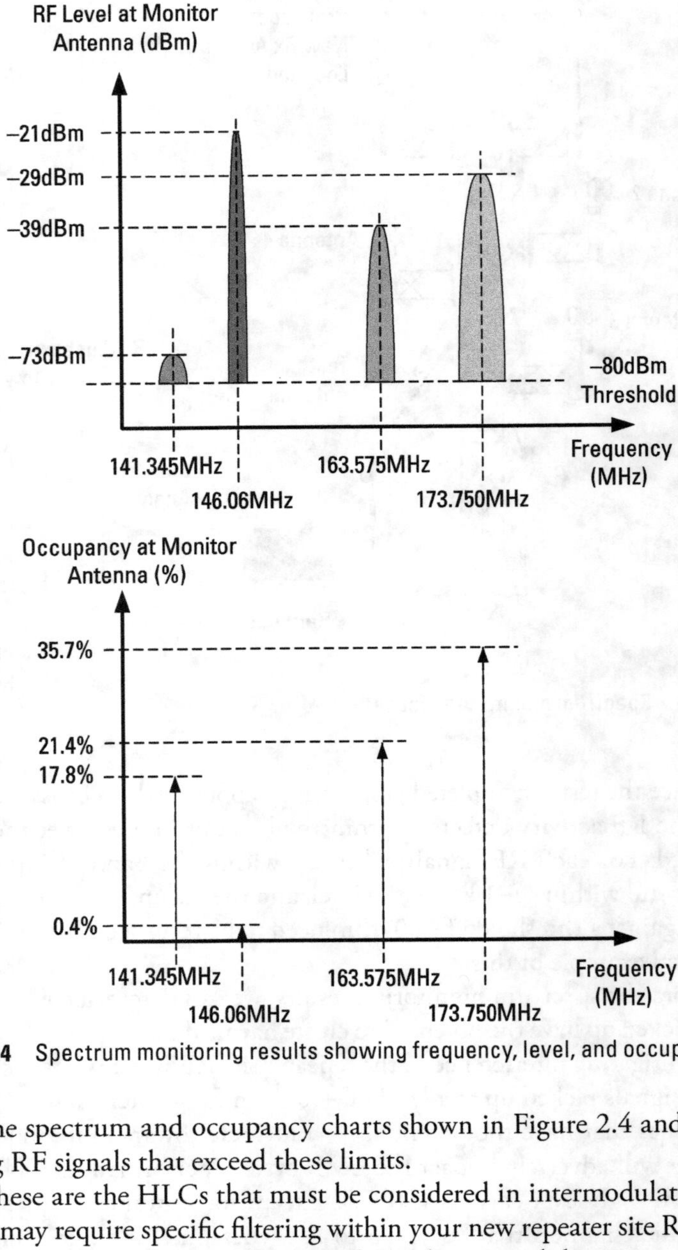

Figure 2.4 Spectrum monitoring results showing frequency, level, and occupancy.

takes the spectrum and occupancy charts shown in Figure 2.4 and lists the existing RF signals that exceed these limits.

These are the HLCs that must be considered in intermodulation studies and may require specific filtering within your new repeater site RF system design to make sure any resulting on-channel intermodulation is suppressed to safe levels; that is, it will not deafen any of the receivers that you plan to deploy there.

Table 2.1
Signal Monitoring Results Showing HLCs

Frequency (MHz)	Average Power (dBm)	Occupancy (%)
163.575	–39	21.4
173.750	–29	35.7

2.3.2 Noise Monitoring

The second part of the RF survey is noise monitoring. This is where we individually monitor the ambient site noise level on every channel over the receive band being considered for the new LMR system. To do this, we add a frequency agile crystal filter to the test system as shown in Figure 2.5 with a bandwidth narrow enough to measure the site noise floor one channel at a time. For LMR systems using 12.5-kHz channels, the bandwidth of the crystal filter will typically be 8 kHz. When combined with a low noise amplifier, it gives us a test system that can measure noise levels very close to thermal.

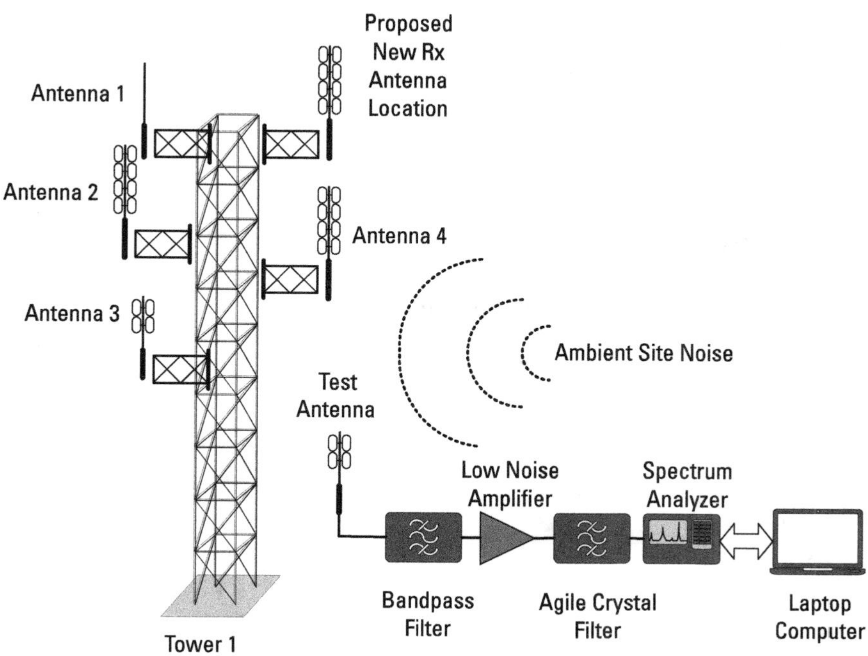

Figure 2.5 Noise monitoring test setup.

The measured results are best provided in Table 2.2, stating the ambient site noise on each channel across the receive band. It is most convenient if this is given both as a discrete level and in terms of dB above thermal (dB above KTB) observed.

Using this, we can choose the best receive frequencies for each individual repeater site in the system, that is, those where the noise floor is low, giving us the greatest uplink (user to site) coverage. For convenience, the site noise data should have the effect of the measurement system removed. This is so that when you come to use the data for designing a proposed new RF system for the site, the measured site noise floor can be added directly (by the rms sum) to the noise floor of the receivers that you plan to deploy. We will see how to do this in Chapter 9.

2.4 Tower Structural Survey

Site surveys give us a general overview of the site, tower mapping tells us where all the existing antennas are, and RF surveys give us a detailed picture of the RF environment. The tower structural survey concentrates on the mechanical condition of the tower on which we are proposing to deploy new antennas. Specifically, it is an analysis of the existing tower as built and designed with conclusions that enable decisions to be made as to if and where further antennas can be added.

Table 2.2
Noise Monitoring Results Showing Site Noise Floor above KTB in Each Channel

	Ambient Site Noise	
Frequency (MHz)	**Measured Noise in 8-kHz Bandwidth (dBm)**	**dB above KTB**
150.0000	−124.2	10.78
150.0125	−123.4	11.55
150.0250	−123.7	11.23
↓	↓	↓
154.9750	−125.5	9.45
154.9875	−125.6	9.27
155.0000	−125.6	9.34

As a minimum, the tower structural survey should provide the following:

- Latitude and longitude of the tower;
- Details of tower, including manufacturer, type, model, height, and ladder location;
- Details of each antenna on the tower: manufacturer, type, model, location, and azimuth;
- Tower mechanical loading calculations resulting from the weight of the antennas on the tower and the transmission lines running to them;
- Details of any antennas already proposed for future installation;
- Drawings of tower with all appurtenances.

In summary, taking the time to carry out these surveys upfront enables the RF system engineer to confidently design repeater site RF systems that will achieve the goals of minimum interference potential and maximum coverage. How to do this using the results of the surveys described is the subject of the design example given in Chapter 12.

3

Essential Background

Before starting on how to design repeater site RF systems, this chapter provides a refresher on noise and intermodulation (IM). This is necessary to understand how they cause interference in LMR communications systems, as covered in Chapter 4, and then how we go about mitigating them in Chapter 5 and the rest of the book. However, for now, it is only necessary to understand what noise and IM are, how they are generated, and how they behave in radio systems.

3.1 Noise in RF Systems

In radio systems, noise is a generic term used to describe the unwanted random fluctuations superimposed on the wanted RF electrical signal. At RF, the noise in a radio receiver is a combination of natural electromagnetic atmospheric noise and man-made RF interference (RFI) picked up by the receiver's antenna, in addition to the thermal, shot, and flicker noise present in the connected receiver's circuitry. Thermal noise is caused by agitation of electrons in a conductor due to temperature. It increases in level with temperature and is a type of white noise because the noise power is the same across the whole radio spectrum used by LMR systems. Shot noise occurs in electrical circuits where there is a potential barrier such as a diode or transistor PN junction. It is caused by electrons randomly crossing back and forth across the junction. Again, shot noise can be considered as white noise at LMR frequencies, but, unlike thermal noise, the level of shot noise is independent of temperature. Lastly, flicker noise occurs at very low levels in most electronic devices for a variety of reasons. It is often known as $1/f$ (pink) noise as the power reduces with increasing frequency.

3.1.1 The Behavior of Noise in RF Systems

Of the types of noise mentioned, shot and flicker noises occur at levels low enough for them to be negligible in radio systems. Of the types of noise that are noticeable, natural electromagnetic noise and man-made RF interference are measurable, but their levels are dependent on the environment. Thermal noise is also significant but occurs at a level according to temperature that can be both measured and calculated.

Thermal noise has a time average value of zero, and a root mean square (rms) voltage of:

$$V_n = \sqrt{(4 \times K \times T \times B \times R)} \tag{3.1}$$

where

V_n = rms thermal noise voltage (V);

K = Boltzmann's constant = 1.38×10^{-23} (J/K);

T = temperature in Kelvin (K = Celsius + 273);

B = bandwidth (Hz);

R = resistance (ohms).

To calculate the thermal noise power present, consider Figure 3.1, which shows the power delivered into a matched load. For a matched load, the rms thermal noise voltage is equally divided between the source resistance R_S and the load resistance R_L. So, the thermal noise power dissipated in the load, P = KTB. If we assumed that the bandwidth, B, was 1 Hz, then it is a simple matter to calculate the noise power at typical room temperature of 290K (+17°C).

$$\begin{aligned} P_L &= K \times T \\ &= (1.38 \times 10^{-23}) \times 290 \\ &= 4 \times 10^{-21}\ \text{W} \end{aligned} \tag{3.2a}$$

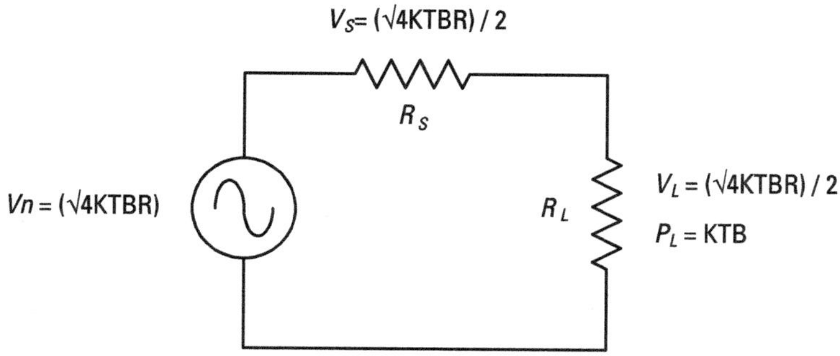

Figure 3.1 Thermal noise power delivered into a matched load.

where

P_L = thermal noise power in the load (watts);
K = Boltzmann's constant = 1.38×10^{-23} (J/K);
T = temperature in Kelvin (K = Celsius + 273).

Converting this to a more convenient unit of dB relative to 1 mW (known as dBm) gives:

$$P_L \text{ (dBm)} = 10 \times \text{LOG}_{10} (4 \times 10^{-21}/1 \text{ mW}) \\ = -174 \text{ dBm/Hz} \tag{3.2b}$$

The unit is dBm/Hz because it is the noise power in dBm in a 1-Hz bandwidth. However, the bandwidth of a practical LMR receiver will be much greater than 1 Hz, so to account for bandwidth, we can rewrite the equation for thermal noise power in the receiver as:

$$P_L \text{ (dBm)} = -174 + [10 \times \text{LOG}_{10} (B)] \tag{3.3a}$$

where

P_L = thermal noise power in the load (dBm);
B = receiver equivalent noise bandwidth (Hz).

Most LMR systems operate in 12.5-kHz channels. However, the LMR signal itself must be narrower so that it fits inside the 12.5-kHz channel bandwidth without spilling too much power into the neighboring channel. Practical LMR receivers therefore have a channel filter bandwidth of around 8 kHz, so applying the bandwidth conversion above, the noise power at the input to such a receiver would be:

$$P_L \text{ (dBm)} = -174 + [10 \times \text{LOG}_{10} (8 \times 10^3)] \\ = -135 \text{ dBm/8 kHz} \quad (3.3b)$$

Note the unit is now dBm/8 kHz because it is the noise power in dBm in an 8-kHz bandwidth. A simple analogy to visualize this, because thermal noise power is constant across the radio spectrum, is the wider the channel filter bandwidth of the receiver, the more noise it lets in.

3.1.2 Noise Figure

Noise figure is a figure of merit that quantifies the degradation in signal-to-noise ratio (SNR) that occurs when a signal is passed through an electronic device or system. In radio systems, it is normally used to quantify the degradation in SNR that occurs when a wanted signal is passed through a receiver.

$$\text{Noise figure (dB)} = \text{SNR at device input} \\ \text{(dB)} - \text{SNR at device output (dB)} \quad (3.4)$$

An illustration of a noise figure is shown in Figure 3.2. Start by assuming that we measured a noise floor of −134 dBm at the input to a low noise RF amplifier using an ideal spectrum analyzer with the resolution bandwidth set to 10 kHz. If we now apply a signal at −100 dBm, the SNR at the amplifier input would be 34 dB. If the amplifier has 15 dB of gain, the signal at the output would increase to a level of −85 dBm. However, if the noise floor measured at the output was −115 dBm, we can see the SNR at the output is now only 30 dB. The noise figure of this amplifier would then be said to be 4 dB because that is how much the SNR has degraded by as the signal passed through the amplifier.

Note that the noise figure can only be used when the input noise to the device is thermal noise only; when additional noise sources are present, these add with the thermal noise at the output of the device.

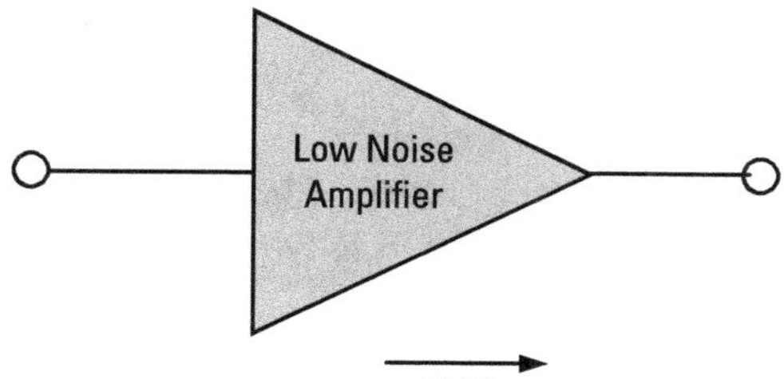

Gain		15dB	
Signal Level (dBm)	−100		−85
Noise Level (dBm/10kHz)	−134		−115
Signal to Noise Ratio (dB)	34		30

Noise Figure (dB) = SNR at Input (dB) − SNR at Output (dB)
= 34 − 30
= 4dB

Figure 3.2 Noise figure.

3.2 IM

When two or more high-level RF signals (F_1 and F_2) are fed into any nonlinear RF device, lower-level harmonics are produced in the device. These mix with each other and the fundamental signals in the nonlinearities to generate IM products, which appear along with the fundamentals at the output, as shown in Figure 3.3.

RF amplifiers and mixers are examples of active RF circuits that produce noticeable IM. However, even some passive devices such as isolators and crystal filters produce IM at levels that can affect radio communication systems.

IM products are classified by their order, which is given by the sum of the coefficients as shown in the following partial list of examples:

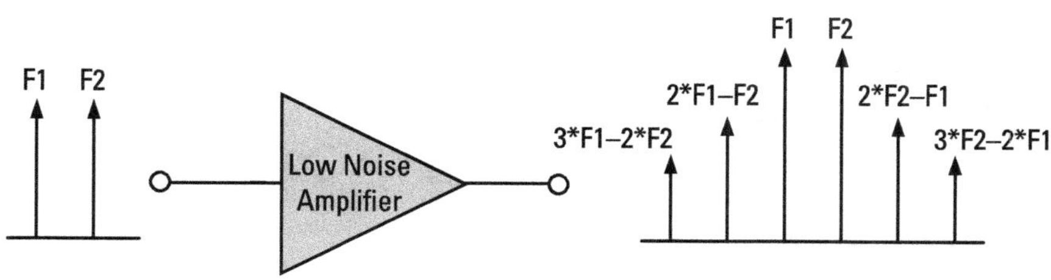

Figure 3.3 IM.

- Second-order IM products: F_1 +/− F_2, F_2 +/− F_1;
- Third-order IM products: $2 \times F_1$ +/− F_2, $2 \times F_2$ +/− F_1;
- Fourth-order IM products: $2 \times F_1$ +/− $2 \times F_2$, $2 \times F_2$ +/− $2 \times F_1$;
- Fifth-order IM products: $3 \times F_1$ +/− $2 \times F_2$, $3 \times F_2$ +/− $2 \times F_1$, and onwards to the sixth order, seventh order, and so forth.

If we calculated the frequencies of the even-order (second-order, fourth-order, sixth-order, and so forth) IM products, we would find they are far removed from the fundamental input signals F_1 and F_2. Conversely, we would find the odd-order (third-order, fifth-order, seventh-order, and so forth) IM products are generated at frequencies very close to the fundamentals, and, for that reason, we are most concerned with these in radio communications systems, as will become clear in Chapter 4 when we discuss how noise and IM actually cause interference in LMR systems.

3.2.1 Third-Order IM and Third-Order Intercept Point

In the same way the noise figure was the figure of merit relating to how much the SNR degraded through a device or system, the intercept point (IP) is the corresponding figure of merit used to describe the linearity. While the IP cannot be directly measured, as it lies well beyond the damage threshold of the device, it can be estimated through measurements made in the device's linear region. Once estimated, it provides a reference from which the levels of IM products can be calculated for any given input or output level of F_1 and F_2.

Considering only the third-order IM products for now, if we were to operate the device in its linear region and increase the levels of the input fundamental signals F_1 and F_2 by 1 dB, we would find that their levels at the output would also increase by 1 dB. However, the levels of the third-order IM products at the output would increase by 3 dB, as shown in Figure 3.4.

If we kept on increasing the levels of F_1 and F_2 at the input, the amplifier would progressively saturate. However, if the fundamental and third-order curves measured in the linear region were extrapolated, we eventually reach a point where their levels would be the same as illustrated in Figure 3.5. This is called the third-order IP (IP3).

The IP3 can be defined at either the input (IIP3) or the output (OIP3) of the device, and basically, the higher the IP3, the more linear the device is and the less third-order IM that it will generate.

Once the IP3 is known, the level of the third-order IM products can be predicted for any level of the input signal. It is a case of working backwards from the IP3. Consider an RF amplifier with an output IP3 of 0 dBm. If the

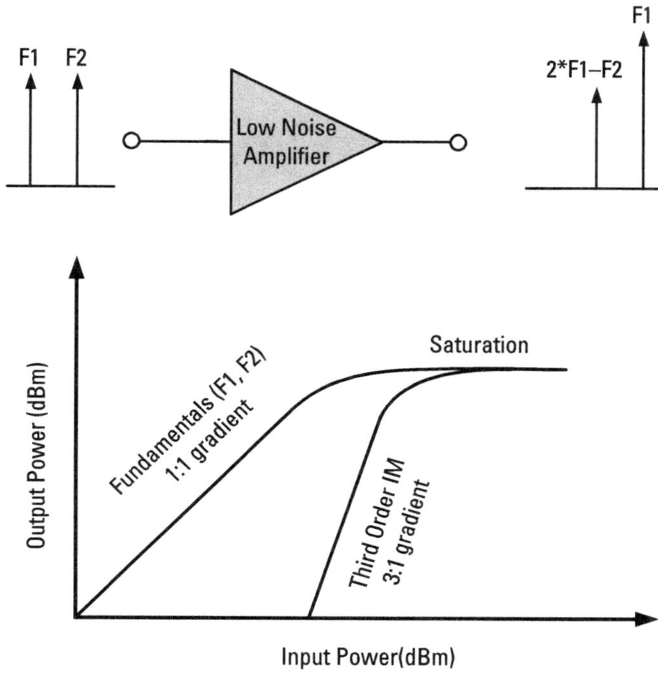

Figure 3.4 Two-tone third-order IM.

levels of the fundamental signals F_1 and F_2 at the output are −10 dBm (i.e., 10 dB below the OIP3), then the levels of the third-order IM products will be 3 times that, or 30 dB below the OIP3. The difference in level at the output between the fundamental signals F_1 and F_2 and the third-order IM products is called the third-order IM rejection ratio (IMR3) and is 20 dB in this example. All this is shown in Figure 3.6.

Mathematically, in this basic case where the levels of the fundamental signals are the same, we can say:

$$\text{OIP3} = (\text{IMR3}/2) + \text{Output Power} \quad (3.5)$$

$$\text{IIP3} = (\text{IMR3}/2) + \text{Input Power} \quad (3.6)$$

$$\text{IMD3} = \text{OIP3} - [3 \times (\text{OIP3} + \text{Output Power})] \quad (3.7)$$

where

OIP3 = output third-order IP (dBm);

IIP3 = input third-order IP (dBm);

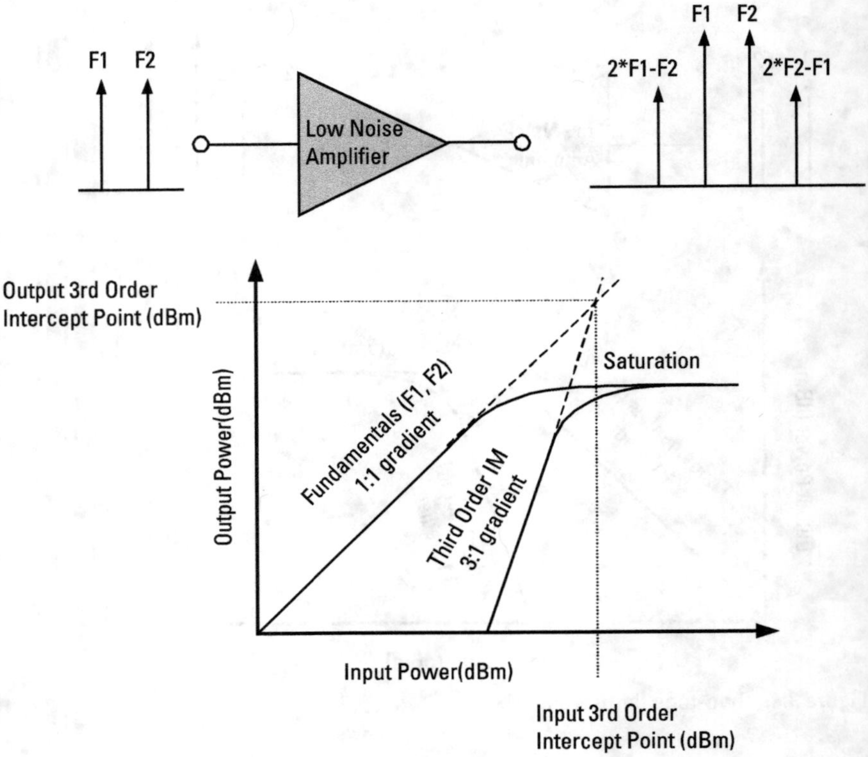

Figure 3.5 Two-tone third-order IP.

IMR3 = third-order IM rejection ratio at output (dB);

IMD3 = level of third-order IM products at output (dBm).

Input power and output power are both in dBm.

3.2.2 Third-Order IM with Unequal Input Signals

Often in practice, though, the levels of the fundamental input signals F_1 and F_2 are not the same. It is when we consider this case that we gain some valuable and more intuitive understanding of how IM actually behaves in RF systems. Let us consider in Figure 3.7 the same RF amplifier with an output IP3 of 0 dBm. The level of F_1 at the output remains at −10 dBm as per the previous example, but the level of F_2 is reduced by a further 10 dB to −20 dBm.

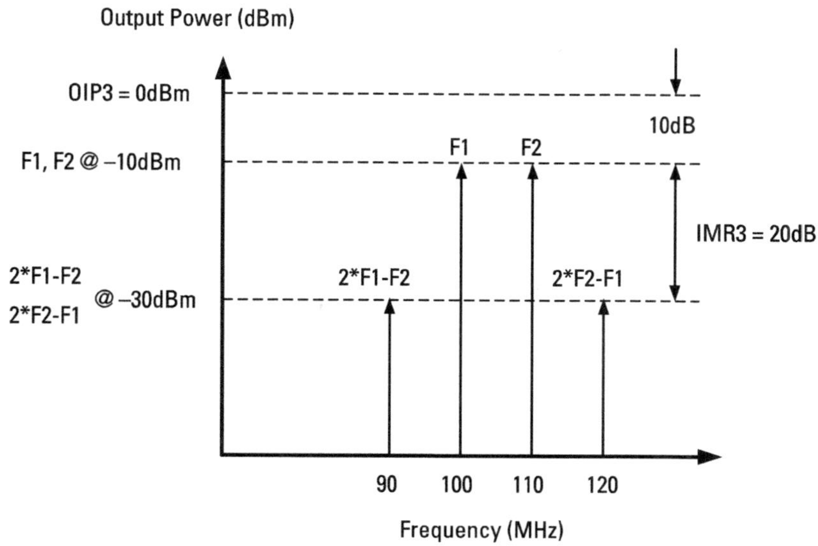

Figure 3.6 Two-tone third-order IM (unequal signals).

Thinking about the level of the $2 \times F_1 - F_2$ product first, notice that the level of F1 is 10 dB below the OIP3. So, because the order of F_1 involved in this IM product is the second harmonic, the level of the $2 \times F_1 - F_2$ product drops by 2×10 dB below the OIP3 to –20 dBm. In addition, the level of F_2 at the output is 20 dB below the OIP3, and because the order of F_2 involved in this IM product is the fundamental, the level of the $2 \times F_1 - F_2$ product drops by another 1 times 20 dB below the OIP3 to –40 dBm. By the same process, the level of the other third-order IM product $2 \times F_2 - F_1$ would be –50 dBm. Figure 3.8 shows what we would see at the output of the device.

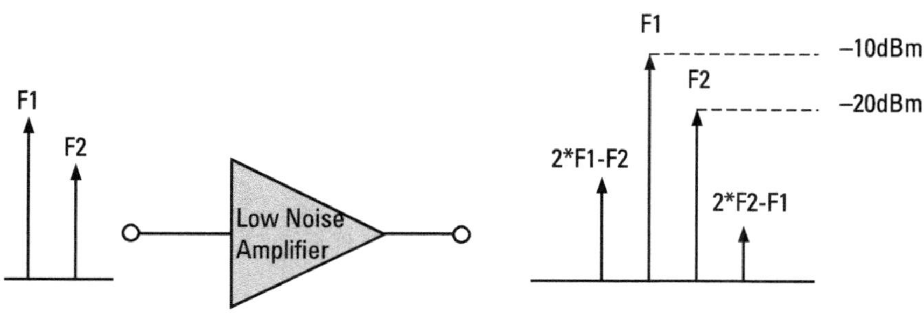

Figure 3.7 Two-tone third-order IM (unequal signals).

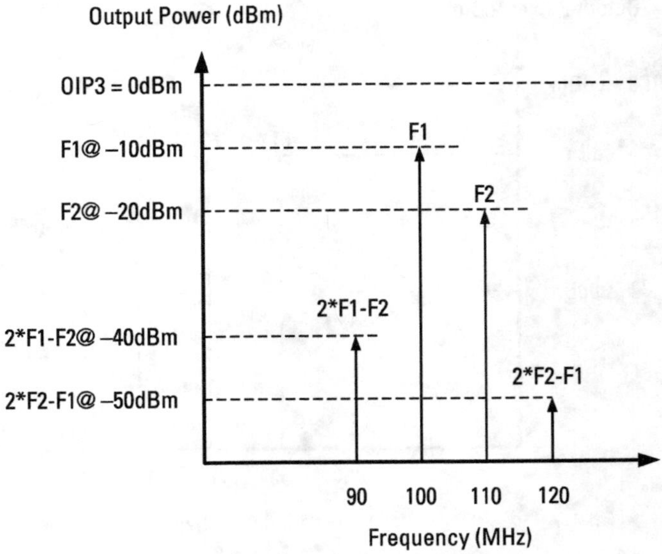

Figure 3.8 Two-tone third-order IM product levels (unequal signals).

The golden rule emerging is the level of the third-order IM products drops according to the order of each frequency involved in the mechanism and how far in the level below the IP3 the fundamental signals F_1 and F_2 are. Thinking back to the first case with equal fundamental signal levels, this gives insight into why, when both signals drop by 1 dB, the IMD3 drops by 3 dB, 2 dB as a result of one of the contributors being the second harmonic (e.g., $2 \times F_1$), and the other 1 dB as a result of the other contributor being the fundamental (e.g., F_2).

In summary then, for two-tone, third-order IM:

- For the $2 \times F_1 - F_2$ mechanism, the IM product level decreases by 2 dB below the OIP3 for every 1-dB drop in F_1 and 1 dB for every 1-dB drop in F_2.
- For the $2 \times F_2 - F_1$ mechanism, the IM level product decreases by 2 dB below the OIP3 for every 1-dB drop in F_2 and 1 dB for every 1-dB drop in F_1.

3.2.3 Third-Order IM with Three Input Signals

So far, we have looked at the behavior of third-order IM products generated by two input signals. Often in practice, though, there will be more than two signals present at the input, so let us now look at what happens if three input signals were applied to the same RF amplifier, resulting in output signals −10 dBm, −20 dBm, and −30 dBm in level, respectively.

Figure 3.9 shows there are now three third-order IM products, but we can use the same empirical rules to estimate their respective levels. Thinking about the level of the $F_1 + F_2 - F_3$ product first, notice that the level of F_1 is 10 dB below the OIP3. So, because the order of F_1 involved in this IM product is the fundamental, the level of the $F_1 + F_2 - F_3$ product drops by 1 times 10 dB below the OIP3 to −10 dBm. In addition, the level of F_2 at the output is 20 dB below the OIP3, and because the order of F_2 involved in this IM product is the fundamental, the level of the $F_1 + F_2 - F_3$ product drops by another 1 times 20 dB below the OIP3 to −30 dBm. Finally, the level of F_3 at the output is 30 dB below the OIP3, and because the order of F_3 involved in this IM product is also the fundamental, the level of the $F_1 + F_2 - F_3$ product drops by another 1 times 30 dB below the OIP3 to −60 dBm. By the same process, the level of the other three-tone third-order IM products $F_1 + F_3 - F_2$ and $F_2 + F_3 - F_1$ would also be −60 dBm, as per Figure 3.10.

In summary then, for three-tone third-order IM, for the $F_1 + F_2 - F_3$, $F_2 + F_3 - F_1$, and $F_1 + F_3 - F_2$ mechanisms, the IM product level decreases by 1 dB below the OIP3 for every 1-dB drop in F_1, F_2, or F_3.

3.2.4 Fifth-Order IM and Fifth-Order IP

Before leaving this subject, it is worth applying the same rules to fifth-order IM, to illustrate what happens there. Starting with the two-tone fifth-order IM products for now, if we increased the levels of the input fundamental signals F_1 and F_2 by 1 dB, we would find that the levels of the fundamental signals at the output would also increase by 1 dB. However, the levels of the fifth-order IM products at the output would increase by 5 dB, as per Figure 3.11.

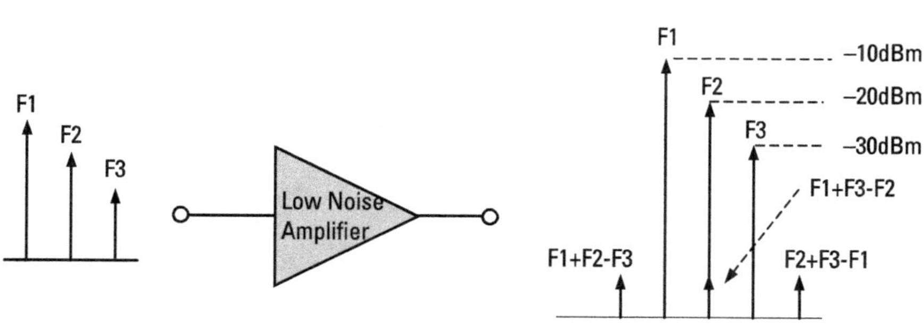

Figure 3.9 Three-tone third-order IM (unequal signals).

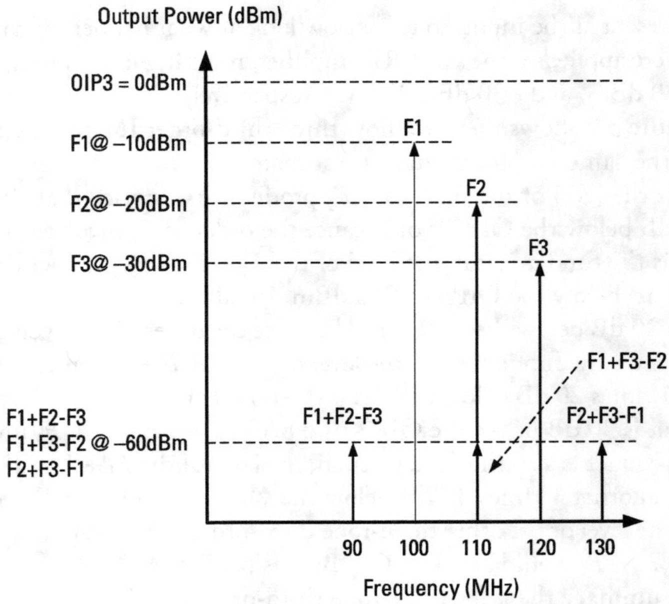

Figure 3.10 Three-tone third-order IM product levels (unequal signals).

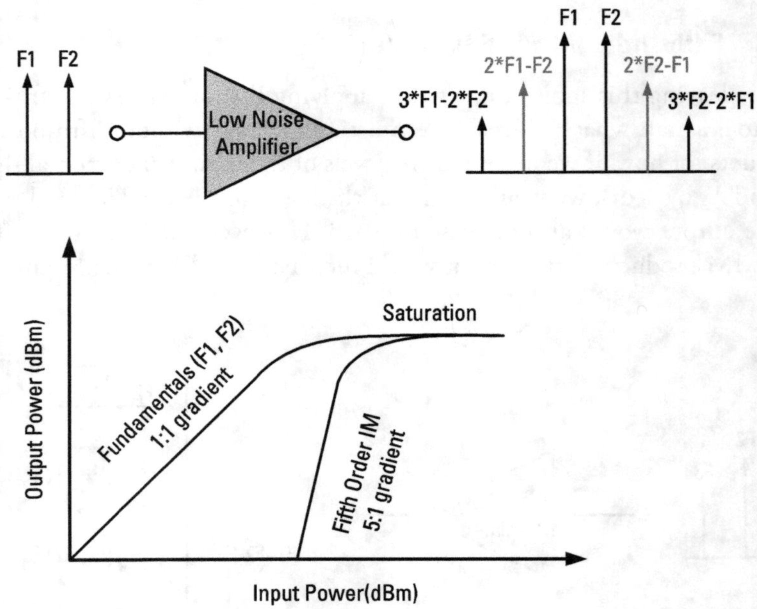

Figure 3.11 Two-tone fifth-order IM.

Increasing the levels of F_1 and F_2 at the input would cause the amplifier to progressively saturate. However, if the linear portions of the fundamental and fifth-order curves were extrapolated, we eventually arrive at the fifth-order IP shown in Figure 3.12 where both the fundamental and fifth-order IM signals at the output would theoretically be at the same level.

Again, the fifth-order IP (IP5) can be defined at either the input (IIP5) or the output (OIP5) of the device, and the higher the IP5, the more linear the device is and the less fifth-order IM it will generate.

Once the fifth-order IP is known, the level of the fifth-order IM products can be predicted for any input signal levels. It is a case of working backwards, but this time from the IP5. Consider an RF amplifier with an output IP5 of 0 dBm. If the levels of the fundamental signals F_1 and F_2 at the output are −10 dBm (i.e., 10 dB below the OIP5), then the levels of the fifth-order IM products will be −50 dBm (or 5 times 10 dB below the OIP5). The difference in level at the output between the fundamental signals F_1 and F_2 and the fifth-order IM products is called the fifth-order IM rejection ratio (IMR5) and is 40 dB in this example. All this is shown in Figure 3.13.

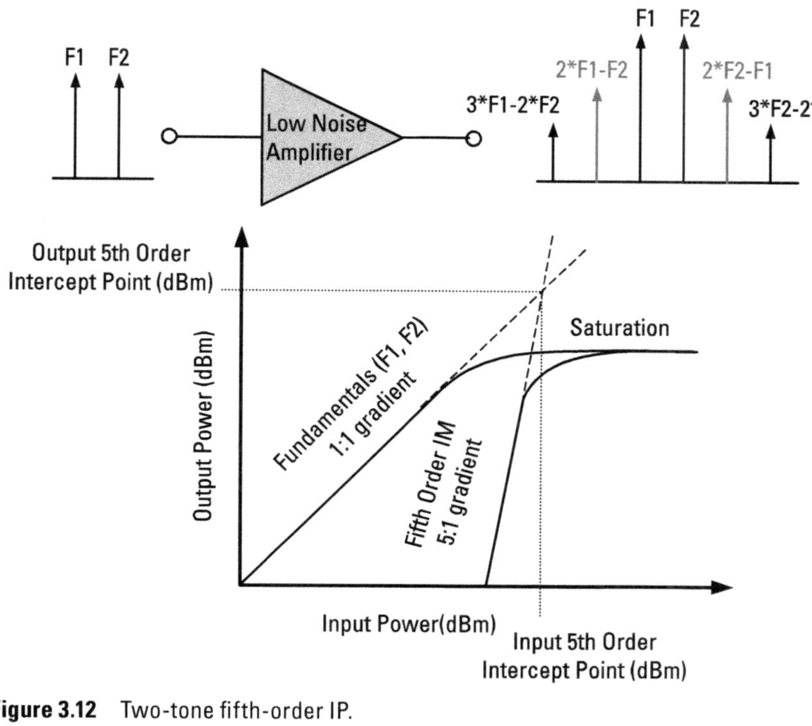

Figure 3.12 Two-tone fifth-order IP.

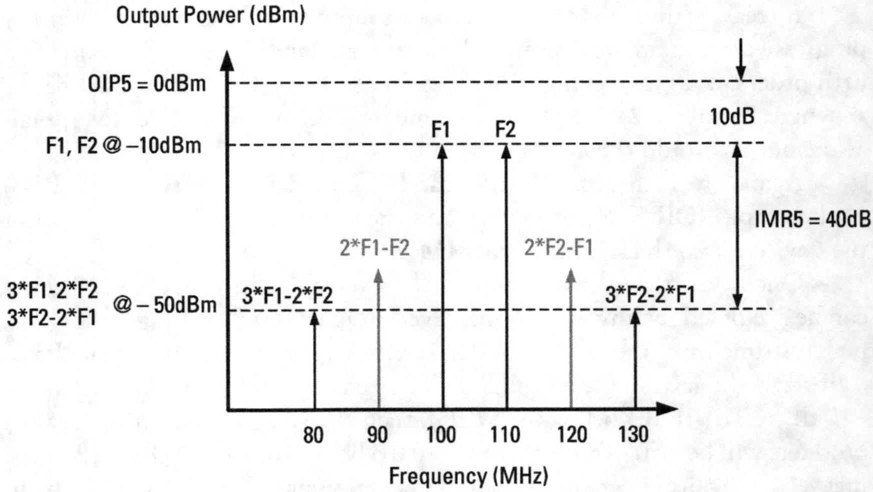

Figure 3.13 Two-tone fifth-order IM product levels (equal signals).

Mathematically, when the levels of the fundamental signals are the same, we can say:

$$\text{OIP5} = (\text{IMR5}/4) + \text{Output Power} \tag{3.8}$$

$$\text{IIP5} = (\text{IMR5}/4) + \text{Input Power} \tag{3.9}$$

$$\text{IMD5} = \text{OIP5} - [5 \times (\text{OIP5} + \text{Output Power})] \tag{3.10}$$

where

OIP5 = Output fifth-order IP (dBm);

IIP5 = Input fifth-order IP (dBm);

IMR5 = Fifth-order IM rejection ratio (dB);

IMD5 = Level of fifth-order IM products at output (dBm);

Input power and output power are both in dBm.

3.2.5 Fifth-Order IM with Unequal Input Signals

Now think about the same RF amplifier with output IP5 of 0 dBm, with the level of F_1 at the output still at −10 dBm, but the level of F_2 reduced by a further 10 dB to −20 dBm, as shown in Figure 3.14.

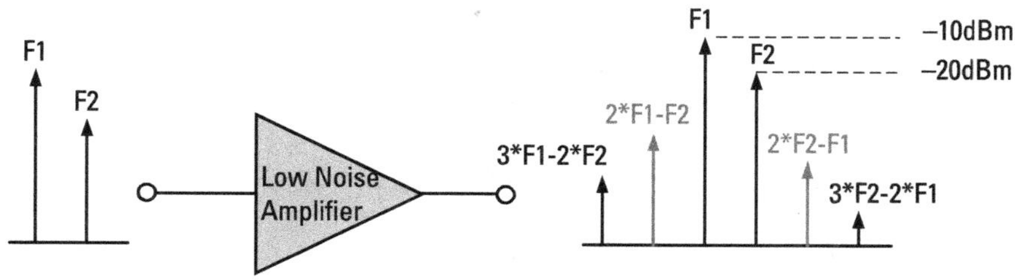

Figure 3.14 Two-tone fifth-order IM (unequal signals).

Thinking about the level of the $3 \times F_1 - 2 \times F_2$ product first, notice the level of F_1 is 10 dB below the OIP5. So, because the order of F_1 involved in this IM product is the third harmonic, the level of the $3 \times F_1 - 2 \times F_2$ product drops by 3 times 10 dB below the OIP5 to −30 dBm. In addition, the level of F_2 at the output is 20 dB below the OIP5, and because the order of F_2 involved in this IM product is the second harmonic, the level of the $3 \times F_1 - 2 \times F_2$ product drops by another 2 times 20 dB below the OIP5 to −70 dBm. By the same process, the level of the other third-order IM product $3 \times F_2 - 2 \times F_1$ would be −80 dBm. Figure 3.15 shows what we would see at the output of the amplifier.

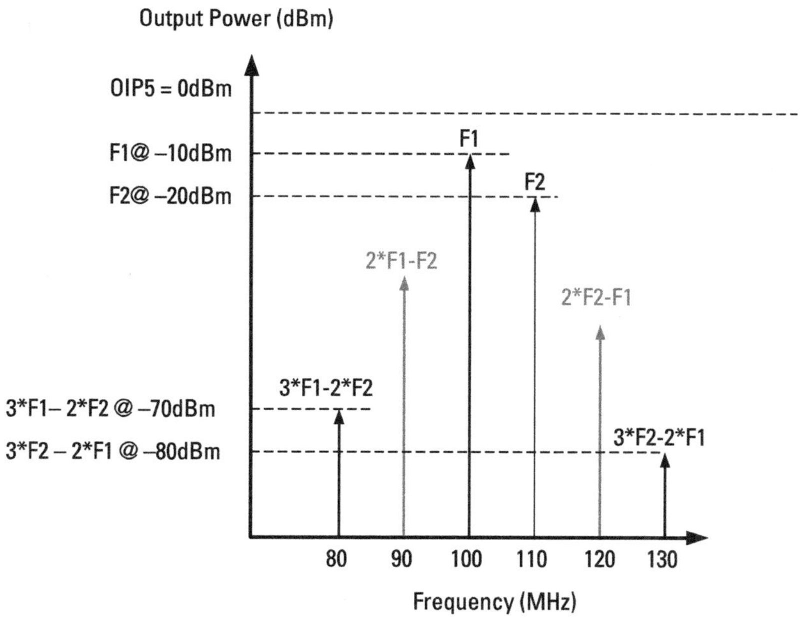

Figure 3.15 Two-tone fifth-order IM product levels (unequal signals).

All we are doing again is applying the rule that the level of the fifth-order IM products drops according to the order of each frequency involved in the mechanism and how far in level below the fifth-order IP the fundamental signals F_1 and F_2 are.

In summary then, for two-tone fifth-order IM:

- For the $3 \times F_1 - 2 \times F_2$ mechanism, the IM level decreases by 3 dB below the OIP5 for every 1-dB drop in F_1 and 2 dB for every 1-dB drop in F_2.
- For the $3 \times F_2 - 2 \times F_1$ mechanism, the IM level decreases by 3 dB below the OIP5 for every 1-dB drop in F_2 and 2 dB for every 1-dB drop in F_1.

3.2.6 Fifth-Order IM with Three Input Signals

Now let us think about what happens if another input signal F_3 is applied to the RF amplifier, so that the output levels of F_1, F_2, and F_3 are –10 dBm, –20 dBm, and –30 dBm, respectively, as shown in Figure 3.16.

Note that there are many more three-tone fifth-order IM products than shown. To keep things deliberately simple as we are only trying to explain concepts here, only two of the nearby three-tone IM5 products are shown in Figure 3.16: $2 \times F_1 + F_2 - 2 \times F_3$ and $2 \times F_3 + F_2 - 2 \times F_1$. We will concentrate on determining the levels of these products in the following text.

Considering the level of the $2 \times F_1 + F_2 - 2 \times F_3$ product first, notice the level of F_1 is 10 dB below the OIP5. So, because the order of F_1 involved in this IM product is the second harmonic, the level of the $2 \times F_1 + F_2 - 2 \times$

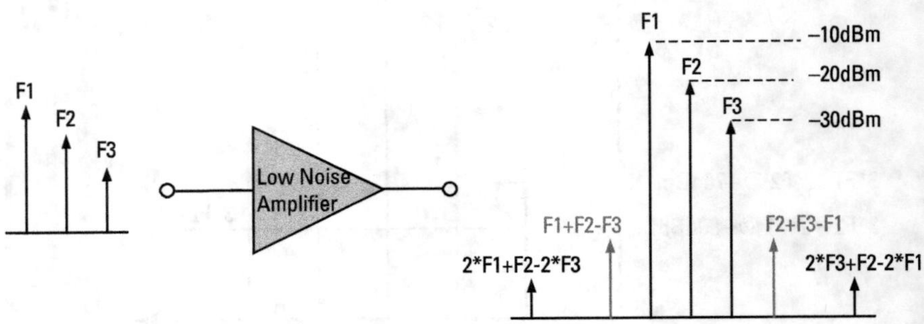

Figure 3.16 Three-tone fifth-order IM (unequal signals).

F_3 product drops by 2 times 10 dB below the OIP5 to −20 dBm. In addition, the level of F_2 at the output is 20 dB below the OIP5, and because the order of F_2 involved in this IM product is the fundamental, the level of the $2 \times F_1 + F_2 - 2 \times F_3$ product drops by another 1 times 20 dB below the OIP5 to −40 dBm. Lastly, the level of F_3 at the output is 30 dB below the OIP5, and because the order of F_3 involved in this IM product is also the second harmonic, the level of the $2 \times F_1 + F_2 - 2 \times F_3$ product drops by another 2 times 30 dB below the OIP5 to −100 dBm. By the same process, the level of the $2 \times F_3 + F_2 - 2 \times F_1$ would be −100 dBm too, as per Figure 3.17.

Note that 0 dBm was used as both the OIP3 and OIP5 in these examples because it is an easy starting point from which to visualize the behavior of IM product levels. In practice, the IP5 of an RF device is almost always lower than the corresponding IP3 by around 5 to 10 dB.

In summary then, for three-tone fifth-order IM:

- For the $2 \times F_1 + F_2 - 2 \times F_3$ mechanism, the IM product level decreases by 2 dB below the OIP3 for every 1-dB drop in F_1, 1 dB for every 1-dB drop in F_2, and 2 dB for every 1-dB drop in F_3.

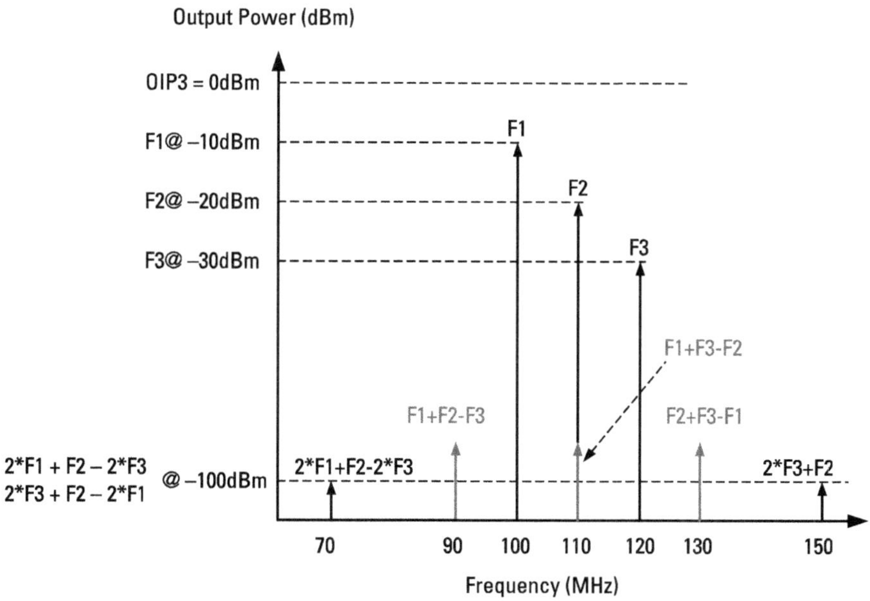

Figure 3.17 Three-tone fifth-order IM product levels (unequal signals).

- For the $2 \times F_3 + F_2 - 2 \times F_1$ mechanism, the IM product level decreases by 2 dB below the OIP3 for every 1-dB drop in F_3, 1 dB for every 1-dB drop in F_2, and 2 dB for every 1-dB drop in F_1.

Finally, it is just worth mentioning the behaviors described here for third-order and fifth-order IM also apply to seventh, ninth, and eleventh-order IM products as well, and well beyond.

4

Sources of Interference at Shared Radio Sites

Interference may be defined as the effect upon the reception of a radio system of one or a combination of transmitted emissions, resulting in serious degradation or loss of communication. Colocating radio repeater equipment at shared sites creates the possibility of interference between the various systems, where RF energy in a receiver passband from nearby transmitters becomes stronger than the intended signal from a distant user. Factors affecting the level of interference at a site include the relative placement of all the transmit and receive antennas, the frequency bands used, the number of active channels, the characteristics of the respective technologies, and the RF performance of the base station equipment.

So far in this book, we have highlighted the dangers posed by interference to users of LMR systems. At shared radio sites, emissions generated by high power transmitters have the potential to deafen sensitive receivers connected to nearby antennas, significantly decreasing the range over which more distant users can communicate using the LMR network. The potentially dangerous situation that can arise is that distant users can hear the nearest site and thus think that they have coverage; however, when they try to set up an important call, they cannot, because the base station receiver at the site is being deafened by local interference and is unable to detect their weak signal.

This is not a situation in which someone who depends on the LMR network for their safety wants to find themselves. Interference often is not constant in nature either, meaning that existing calls on the LMR network can suddenly be dropped for no apparent reason when a transmitter connected to a nearby antenna becomes active.

Note that it is not just the transmitters at the same shared site that are capable of contributing to interference; high power transmitters at neighboring sites (especially those with high gain directional antennas) can also be involved and may well need to be considered in interference analysis. That is another key value of the RF surveys described in Chapter 2, as such monitoring will reveal all high-level signals present at the receive antenna, whether they originate from that same site or not.

Getting to the bottom of, and resolving, interference issues at shared radio sites is a tedious, time-consuming process, so, when deploying a new RF system on an existing site, it is better to mitigate interference upfront by careful design. This does not just mean designing your system so that your receivers are protected from interference resulting from colocated transmit antennas; it also means preventing your transmitters from interfering with other receivers too. The first step in doing this is to understand the various ways that site RF transmitters can generate interference and deafen one or more receivers connected to nearby antennas. That is what this chapter is all about.

4.1 Transmitter Harmonics

Transmitters generate harmonics at integer multiples of their transmit frequencies. For example, a transmitter operating at 400 MHz will generate harmonics at 800 MHz (second harmonic), 1,200 MHz (third harmonic), 1,600 MHz (fourth harmonic), and so forth. Generally, but not always, the level of the harmonic energy decreases as the order of the harmonic increases; that is, the level of the second-harmonic emission will be the highest, followed by the third harmonic, and so forth. The bandwidth of all the harmonic emissions also increases in a similar way; the bandwidth of the second-harmonic emission will be twice that of the fundamental, the third-harmonic bandwidth will be three times that of the fundamental, and so on. Typically, for an LMR system, the bandwidth of the transmitted signal will be around 8 kHz, so the bandwidth of the second harmonic emission will be 16 kHz, and the bandwidth of the third-harmonic emission will be 24 kHz, and so forth. As LMR systems usually operate in 12.5-kHz channels, this means that harmonic emissions from one LMR transmitter can fall across several LMR receive channels. Note that

Figure 4.1 is illustrative only; the respective bandwidths of the harmonics are not to scale but simply emphasize the point that harmonic energy spreads.

If harmonic energy produced by a transmitter happens to fall within the pass bandwidth of a receiver located at the same site, then it is likely that receiver will be deafened to some extent. The European regulatory body European Telecommunications Standards Institute (ETSI) mandates that all unwanted emissions from a radio transmitter must be at a level of −36 dBm or below. In the United States, the Federal Communications Commission (FCC) mandated limit is less stringent at −20 dBm. For users located well away from the radio site, emissions at this level are not a problem, but for colocated receivers at the site, signals of this strength can easily cause interference if some of the energy falls within the pass bandwidth of a receiver connected to a nearby antenna.

Another way to visualize it is that most LMR receivers at radio sites can detect signals from users in the field down to a level of around −110 dBm (we will see examples of this in Chapter 6 and its following chapters). So, any transmitter that radiates a harmonic at −20 dBm will interfere with a colocated receiver operating at that frequency if the isolation between the transmit antenna and that of the affected receiver is less than 105 dB (because, to be detected, the signal needs to be about 15 dB above the noise and interference,

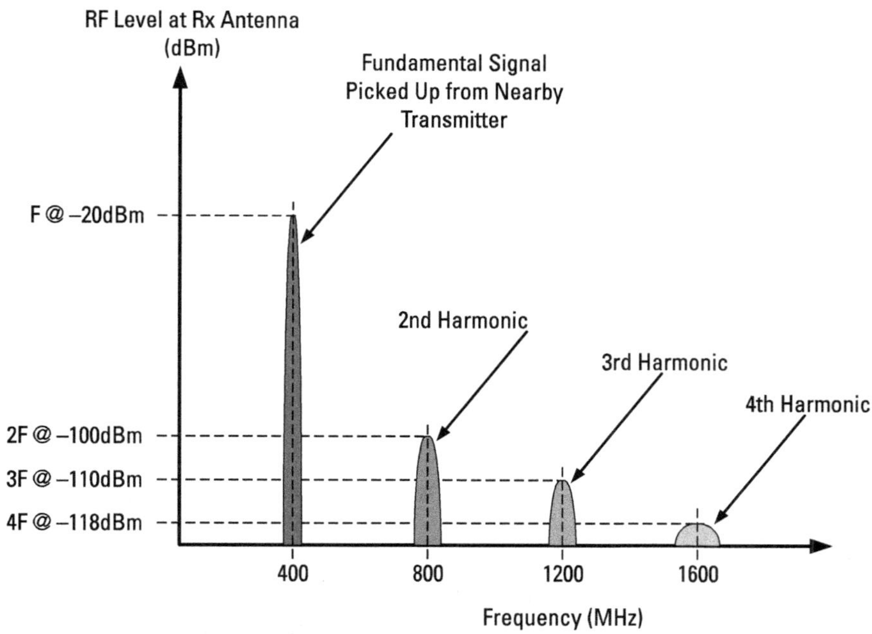

Figure 4.1 Transmitted signal and harmonics seen at the receive antenna.

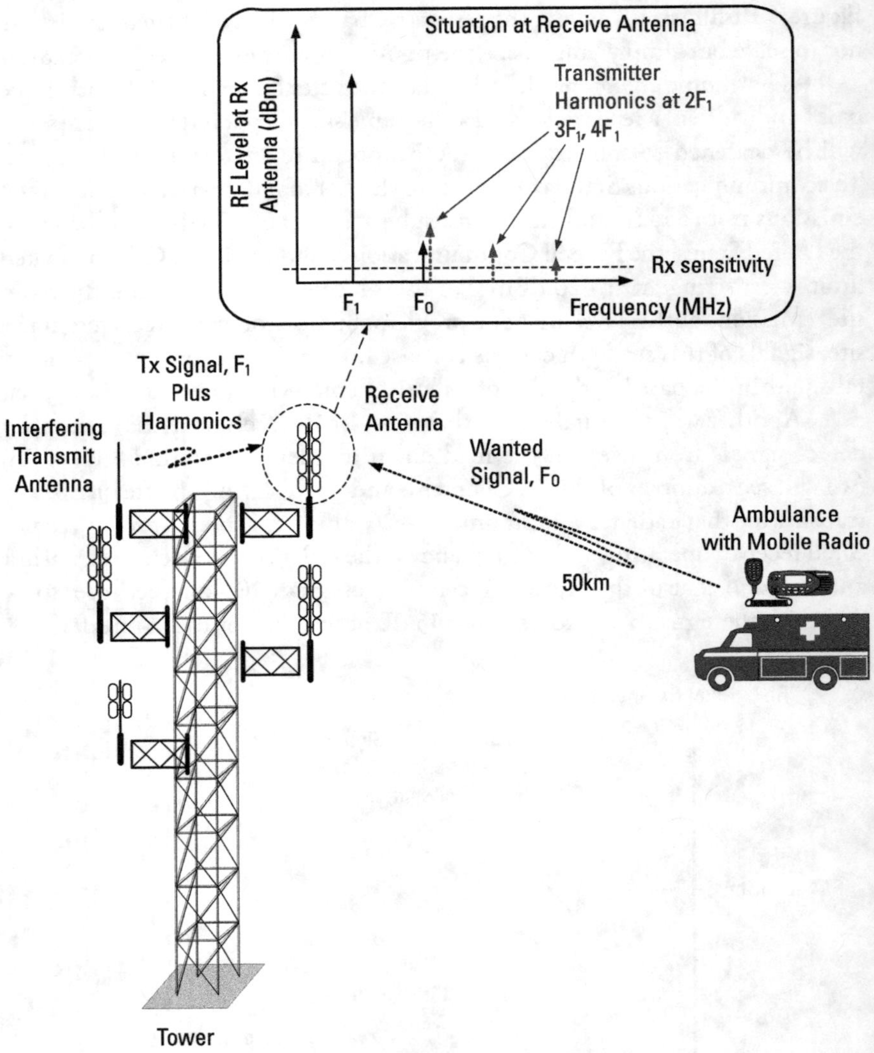

Figure 4.2 Second-harmonic interference at the receive antenna.

as we will see in Chapter 6). Figure 4.2 shows how a second-harmonic emission from a transmitter can deafen a receiver operating at the second harmonic frequency. In the diagram, F_1 is the transmit signal from a colocated antenna on the tower, and its second harmonic $2 \times F_1$ happens to sit right at the receive frequency of the ambulance system, F_O. The result is that the transmission from the ambulance is lost.

In summary, because of the high isolation required to prevent harmonic interference, it is better to avoid deploying new receivers at a shared radio site that uses frequencies that are exact harmonics of existing transmit frequencies.

4.2 Transmitter Sideband Noise

Any electrical signal contains unwanted random fluctuations, broadly termed "noise." Applied to radio, this means that the output from RF transmitters is not totally clean as shown in Figure 4.3; it also contains much lower but detectable amounts of energy (noise) at frequencies well above and below the desired transmit frequency.

On shared radio sites, the danger is this unwanted energy (known as transmitter sideband noise) can be picked up on nearby receive antennas and deafen any connected receivers, especially those operating close to the transmitter frequencies. As the level of transmitter sideband noise decreases the further you go away from the transmit signal, the most important frequency at which to measure the noise is that of the closest site receive frequency. At

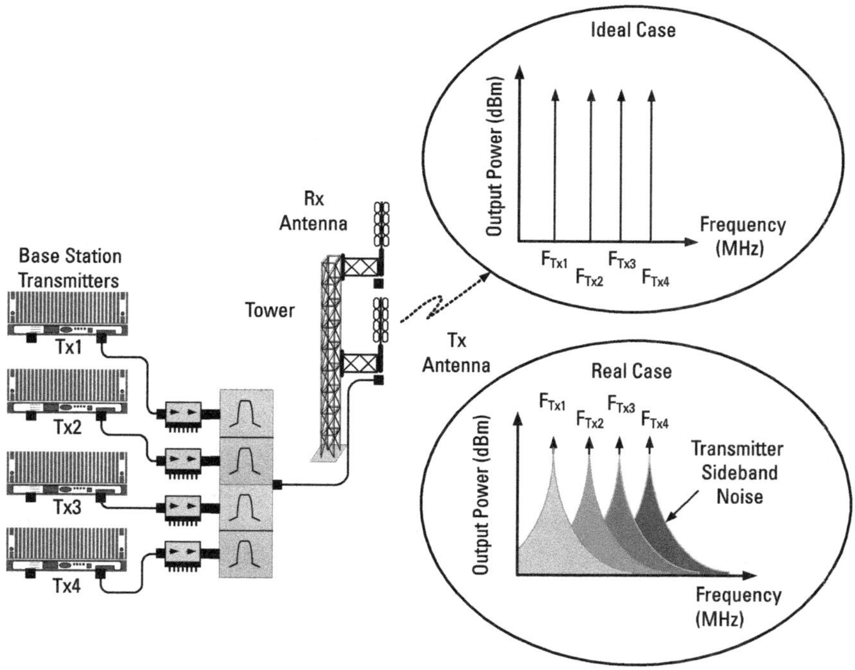

Figure 4.3 Transmitter sideband noise.

the frequencies used by LMR systems, there are generally fixed offsets between the site transmit and receive frequencies, usually 4.6 MHz for very high frequency (VHF) systems, 9.5 MHz for ultrahigh frequency (UHF) systems, and 45 MHz for 800-MHz systems, so we are usually most interested in the sideband noise at these offsets, because it corresponds to the frequencies in which related site receivers are listening.

Figure 4.4 shows that the weak received signal from the distant ambulance, F_O, is still above the site receiver's sensitivity level and again would normally be detected. However, the transmit antenna directly below on the tower is radiating sideband noise that falls within the receiver's bandwidth at a level preventing the weak signal from the ambulance being heard. A simple analogy is to think of two people standing at opposite ends of a large room. Usually, they would be able to speak to each other and communicate, but if a constant noise is played directly into the ear of one of them, that person now has trouble hearing the other person.

The mitigation of this problem involves designing the filtering in the transmitter combiner so that any sideband noise at the nearest site receive frequency is suppressed well below the receiver's sensitivity. How to do this will be addressed in detail in Chapter 8.

Finally, it is just worth mentioning that the level of transmitter sideband noise at a given offset from the transmit frequency is usually quoted as a normalized discrete level in dBm/Hz or a level relative to the carrier level, in dBc/Hz. Normalizing to a 1-Hz bandwidth is convenient because, by adding to this a factor equal to 10 times the logarithm of the affected receiver's bandwidth (in Hz), we can quickly estimate the total noise power that receiver experiences.

4.3 Transmitter IM

As highlighted in Chapter 1, it is impractical for each transmitter on a shared radio site to have its own dedicated antenna; we would soon run out of real estate on the radio towers. Instead, the approach taken in repeater site RF systems is to combine a number of transmitter outputs together on to a common antenna. This may save real estate, but, as with any system where a number of different signals are present, mixing can occur in any nonlinearities and generate unwanted IM products. Particular nonlinearities within site transmitter combining systems where IM can be generated are:

- The transmitter outputs, due to a signal from a neighboring transmitter feeding round via the combining network as shown in Figure 4.5;

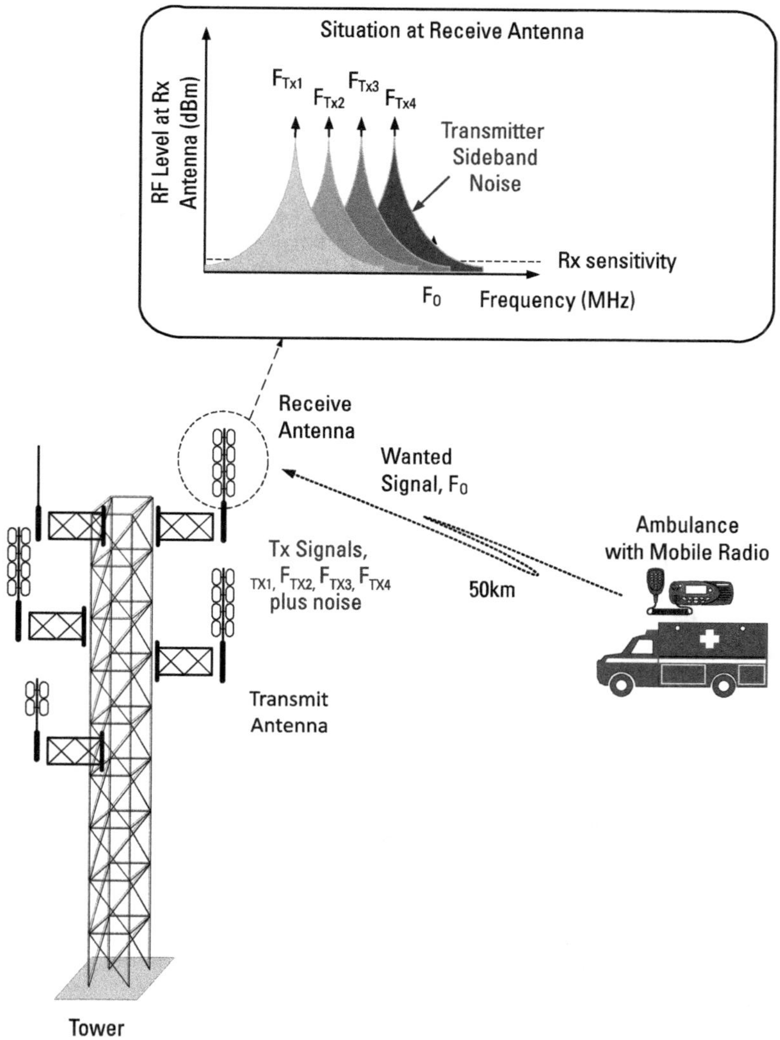

Figure 4.4 Interference caused by transmitter sideband noise.

- The ferrite materials used in the isolators;
- Damaged cables and connectors in the antenna systems connected to the combiner output.

Interference occurs when transmitter IM products picked up by nearby receive antennas at the site happen to coincide with the operating frequency

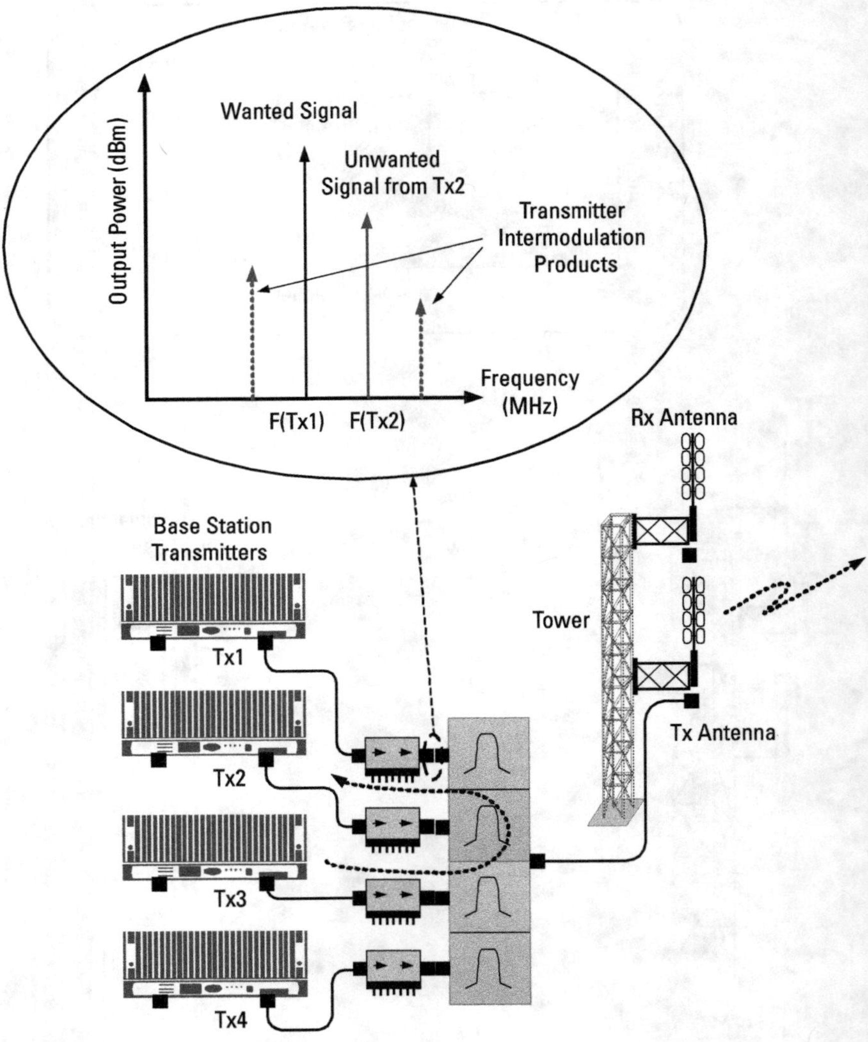

Figure 4.5 Transmitter IM.

of a connected receiver. While the level of the IM produced in the transmitter outputs or the isolators is attenuated by the bandpass filters in the transmitter combiner, it is still possible to radiate enough energy to deafen affected receivers.

Figure 4.6 shows this again using our example of a weak received signal from the distant ambulance, F_O. The signal received at the site is still above

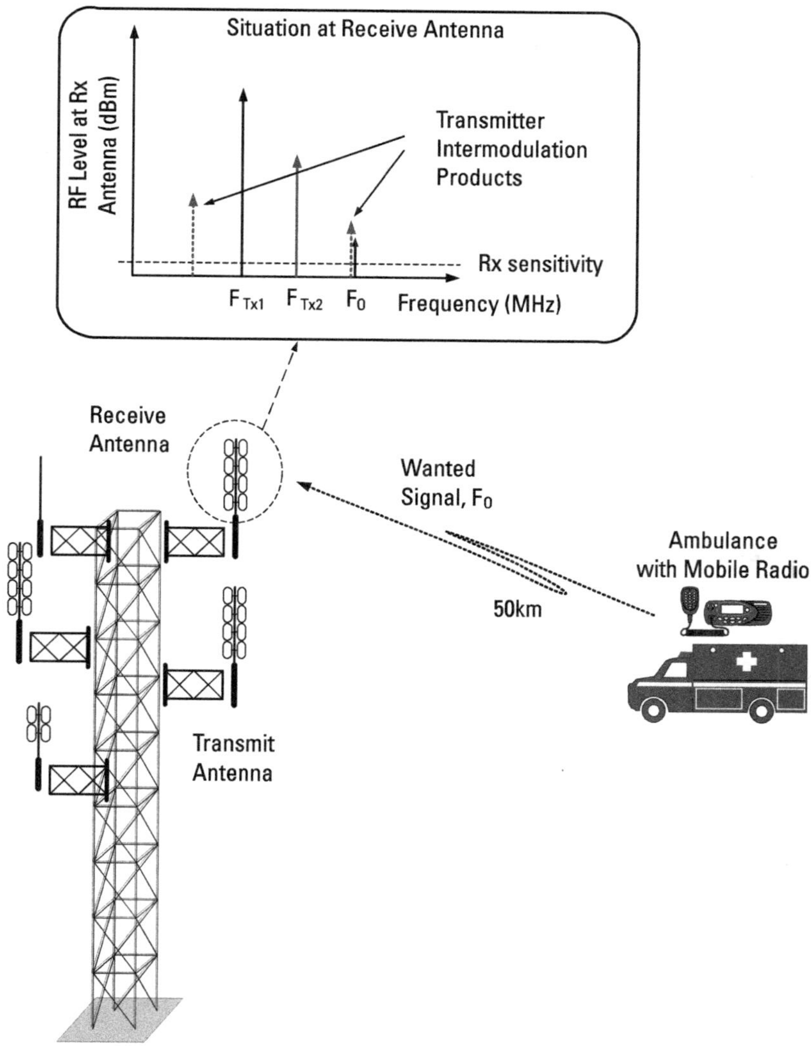

Figure 4.6 Interference caused by transmitter IM.

the site receiver's sensitivity level and should be detected. However, within the transmitter combiner feeding the antenna below, two transmit signals, F_{TX1} and F_{TX2}, are combined and an IM product close to F_O is generated at a level preventing the weak signal from the ambulance being heard.

The mitigation of this problem is best achieved by not combining transmitters that could generate an IM product at a site receive frequency. If unavoidable, then the level of the IM products must be suppressed within the transmitter combiner by means of filtering and isolation between the transmitters. How to do this will again be addressed in detail in Chapter 8.

4.4 Receiver Blocking

The interference mechanisms covered so far, harmonics, sideband noise, and transmitter IM, all have to be suppressed within the site transmitter combining systems to levels that do not cause interference (how to calculate this level is the subject of Chapter 6). Once they are propagated by the Tx antenna, any radiated RF energy that falls within a colocated site receiver's passband cannot be addressed further in the receiver multicoupler filtering, as it is effectively on-channel.

The final two mechanisms that we will examine, blocking and receiver IM, are the opposite case. These are the result of strong off-channel signals picked up by the receive antenna, and so attenuation of these interferers must be carried out within the receiver multicoupler filtering.

Receiver blocking occurs when very strong RF emissions from unwanted sources are picked up on a receive antenna and affect the receiver's circuitry in such a way that it results in a reduction in sensitivity. In general, receiver blocking occurs in one of two ways within the front end (RF) circuitry of the affected receiver:

- *Low noise amplifier overload:* where the strong off-channel interfering signal drives the low noise amplifier in the receiver front end in to gain compression;
- *Reciprocal mixing:* where the strong off-channel interfering signal passes through the receiver front end filters to the mixer RF port and transposes local oscillator noise on to the receiver intermediate frequency (IF) circuitry.

Receiver blocking most commonly happens at two places in an LMR system:

- At the shared radio sites, where strong RF emissions from nearby transmit antennas at the site are picked up and prevent reception of a weak signal from a distant user. This is shown in Figure 4.7.
- In the system's fringe coverage areas far away from the radio site. There, strong emissions from a nearby source (e.g., another radio

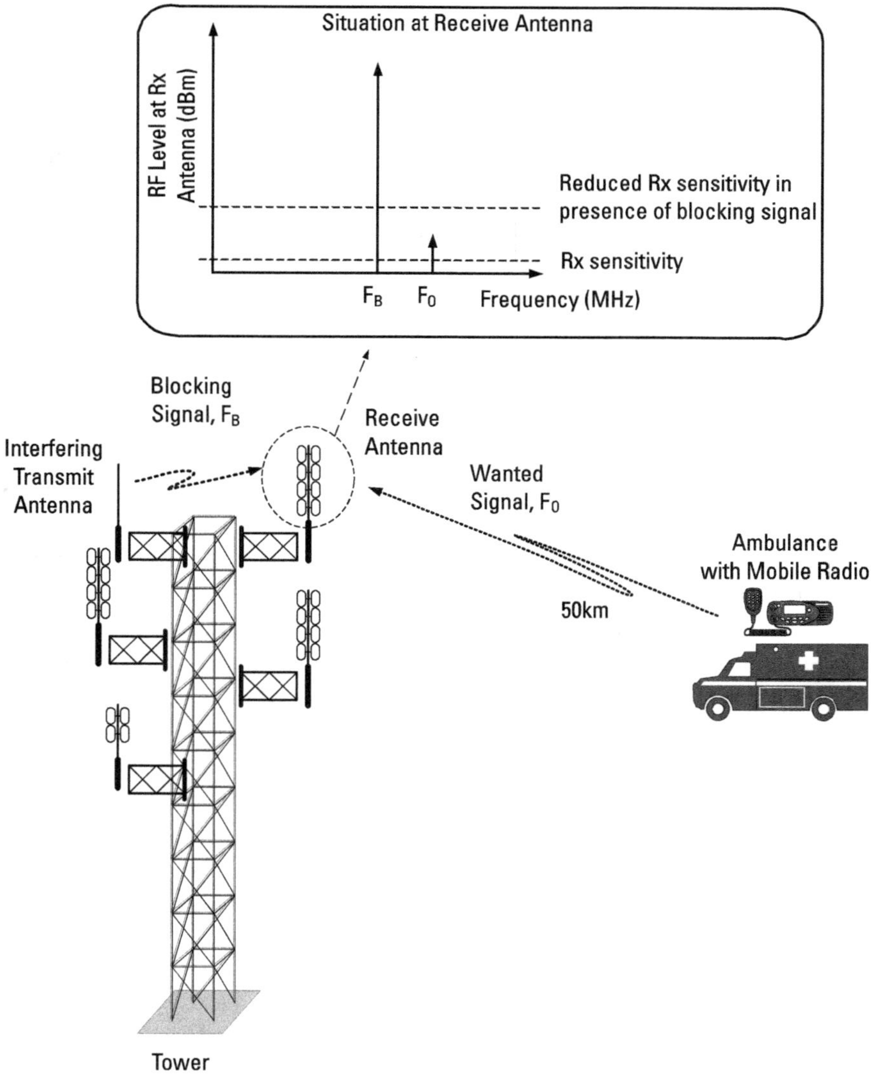

Figure 4.7 Receiver blocking.

site) effectively reduce the sensitivity of the user's mobile or portable receiver and prevent them receiving the weak wanted signal from the distant site to which they are registered.

The weak received signal from the distant ambulance is still above the site receiver's sensitivity level. However, in the presence of the blocking signal from the nearby transmitter, F_B, the receiver circuitry is overloaded and the sensitivity reduces to the point where the signal from the ambulance can no longer be heard. Again to use a simple analogy, think of two people standing at opposite ends of a large room and communicating by speaking to each other. Now, if another person (the blocking transmitter) speaks directly into the ear of one of them, that person now cannot hear what the other person is saying.

Intuitively, many receiver blocking problems can be avoided by sensible site planning, essentially by keeping the antennas of high power transmitters away from those of sensitive receivers using adjacent frequencies, and by using base station radio equipment with good receiver blocking performance.

4.5 Receiver IM

Receiver IM is a measure of the level of unwanted IM products generated in the site receiving system (receive antenna, receive antenna feeder cable, receiver filter, receiver multicoupler, and connected base station receivers). It differs from transmitter IM in this way; transmitter IM is generated in the transmitter combiner and involves only the frequencies of the connected transmitters. Receiver IM is produced within the passive (e.g., antennas and cables) and active (e.g., amplifiers and mixers) nonlinearities of the site receiving system and is a result of the frequencies of all the high-level signals to which the receive antenna is exposed, as shown in Figure 4.8.

The weak received signal from the distant ambulance, F_O, is still above the site receiver's sensitivity level. However, in the presence of the high-level interfering signals, F_1 and F_2 from the nearby transmitters, the site receiver system produces IM that falls within the receiver's bandwidth at a level meaning that the weak signal from the ambulance can no longer be heard. Again to use a simple analogy, think of two people standing at opposite ends of a large room and communicating by speaking to each other. Now, if two more people (the interfering transmitters) start speaking directly into the ears of one of them, that person now cannot hear what the other person is saying.

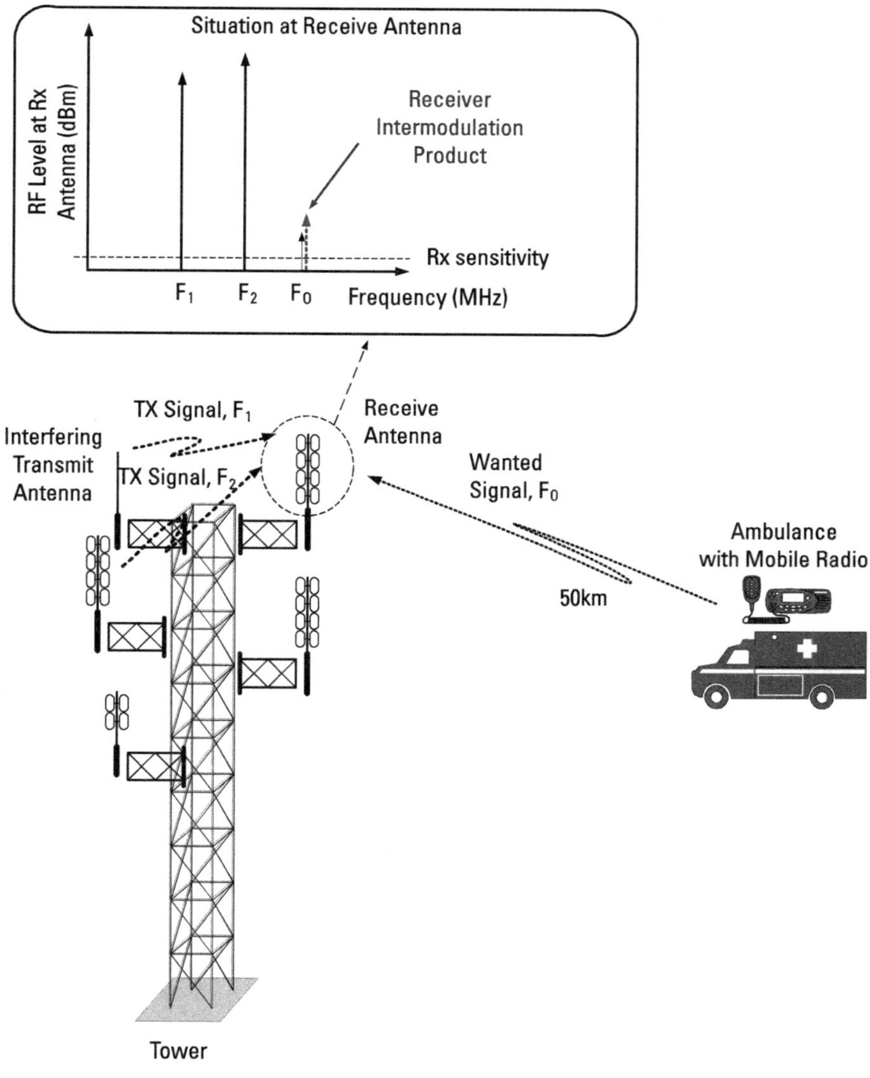

Figure 4.8 Receiver IM.

The mitigation of this problem involves understanding what high-level signals the receive antenna will be exposed to and then designing the filtering in the receiver multicoupler so that any in-band IM products are suppressed well below the receiver's sensitivity. How to do this will be addressed in detail in Chapter 9.

4.6 Minimizing Site Interference by Antenna Placement and Frequency Planning

This section is really important and deserves your undivided attention. Two of the main tools at the RF system design engineer's disposal to minimize interference at a shared radio site are antenna placement and frequency planning. Good antenna placement involves maximizing the isolation between the various transmit and receive antennas at the site. Identifying susceptible receivers and maximizing isolation between the interfering transmit antennas and the potential victim receiver's antenna will minimize the size, cost, and losses of the filtering required (in the various transmitter combiner or receiver multicoupler systems) to mitigate the interference. The rules for best-practice antenna placement when deploying a new system at a shared radio site are covered in Chapter 5. However, even armed with those rules, you may be faced with a situation at well-populated sites where the only positions available for your antennas on the tower are the hitherto unwanted locations. That is when good frequency planning becomes doubly important.

Frequency planning is first carried out at an overall system level to optimize the performance of the network. The general aim is to minimize cochannel and adjacent channel interference that occurs in the desired service area because of overlap in the coverage footprints of the various radio sites. However, the implications of the overall system frequency plan then need to be considered at a site level, specifically at each of the radio sites individually. General good practice rules such as maximizing the separation between transmit and receive frequencies and not combining transmitters that will generate IM at site receive frequencies should always be followed. However, once site surveys, tower mapping, and RF surveys have been carried out, we will know the transmit and receive frequencies of the existing systems operating at each site, and we will know where the respective antennas are. Then it is possible to calculate the frequencies and levels of each potential interfering signal at the location of any new proposed receive antenna and get an appreciation of which of the interference mechanisms described in this chapter may be at play. Many of these, such as IM caused by a combination of the proposed new systems transmit frequencies and existing ones, can often be avoided by simply changing the frequencies of the channels planned for deployment at the site. The overall frequency plan can then be iterated to include these site-specific changes, often at very little (if any) cost to overall system performance.

5

Antennas and Site Antenna Systems

After a detailed site survey and any necessary spectrum and noise monitoring have been carried out and the frequency planning completed, work on designing the actual site RF system can begin. The first and most important step of this is the design of the antenna system, basically choosing the right types of antennas to use and determining the best locations on the site to put them. It sounds simple, but, in reality, it is anything but; because in the majority of situations, you are sharing the tower with many existing antennas connected to a number of other radio transmitters and receivers. Placing your receive antenna in a location that minimizes the number and level of interfering signals to which it is exposed is not a trivial task, and neither is placing your transmit antenna so that your transmissions do not deafen yours or anyone else's receivers.

Get the antenna system design right and the benefits will be maximum coverage, minimum interference, and cost-effective transmitter combining and receiver multicoupling systems where losses, physical size, and cost are minimized. Get it wrong and the cost, losses, and physical size of the transmitter combining and receiver multicoupling systems will all increase along with the probability of interference resulting in reduced coverage. This is potentially dangerous as distant users think they have coverage because they can hear the site, whereas, in reality, they cannot make a call on the network as the weak remains of their transmission is drowned out by the interference at the site.

What an antenna basically does at the transmit end is convert the applied RF voltage into an electromagnetic field that propagates out over the surrounding area. The strength of this electromagnetic field decreases in a manner inversely proportional to the square of the distance the further away you go from the site, so that, at the receiving end, the antenna serves to convert the minute remains of the electromagnetic field into a very small RF electrical voltage that can be detected by the receiver circuitry.

It is not necessary for us to design our own antennas as they can be purchased from several reputable vendors. Instead, this chapter concentrates on the main specifications used to describe antenna performance and then looks at some common types of antennas used at radio repeater sites. Understanding these and the basics of how they work will help you to make the right choice of antenna for each site in your LMR system. The chapter will conclude with some simple design rules, which, if followed, will result in the design of an optimal site antenna system that simplifies interference mitigation and supports the goal of maximum coverage.

5.1 The Isotropic Radiator

An isotropic radiator is a hypothetical antenna that radiates electromagnetic energy equally in all directions. The Sun is the most obvious example of an isotropic radiator, but, in reality, it is not practical to build such an antenna. Despite this, an isotropic radiator is useful as a reference against which to quantify the performance of practical antennas. For example, the gain of an isotropic radiator is said to be 0 dBi (dB relative to isotropic).

Polar plots, with the antenna located at the origin, are often used to illustrate the radiation pattern of an antenna. For an isotropic radiator, the radiation pattern would be like that shown in Figure 5.1 whether viewed from above or the side.

If viewed in 3 dimensions (3D), the radiation pattern would look spherical.

5.2 Antenna Specifications

5.2.1 Directivity

The directivity of an antenna is defined as how many times greater the power radiated in the direction of the strongest emission is compared to the average power radiated in all directions.

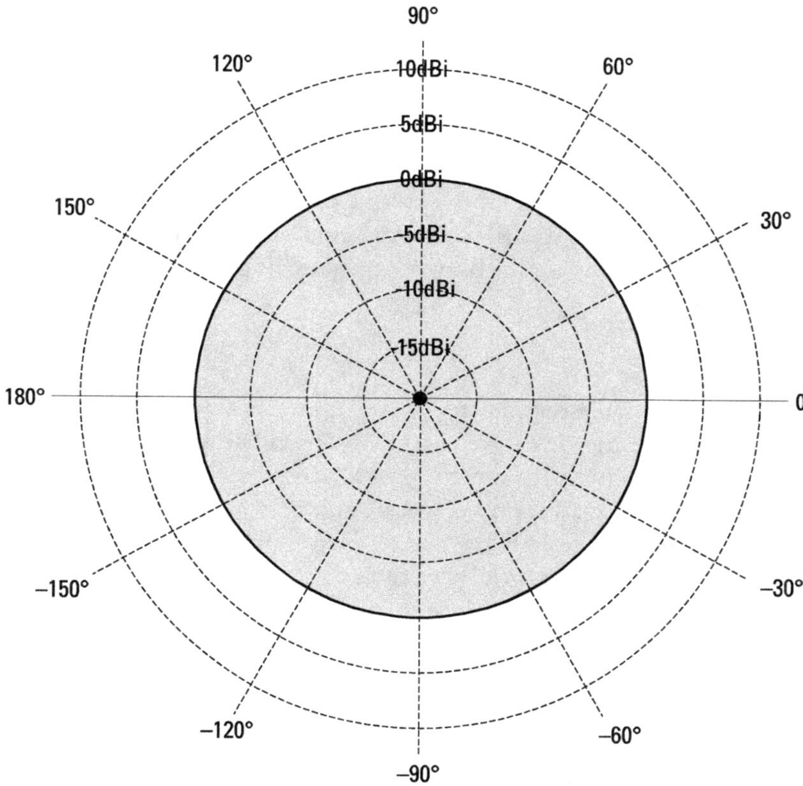

Figure 5.1 Polar plot for an isotropic radiator.

$$\text{Directivity (dB)} = 10 \times \text{LOG}_{10}$$
$$\text{(Maximum radiated power/Average radiated power)}$$

Directivity is an important measure because most emissions from an antenna are intended to go in a particular direction or at least in a particular plane; emissions in other directions or planes are wasteful (or even worse, interfere with other nearby systems).

The directivity of an actual antenna can vary from 1.76 dB for a short dipole to as much as 50 dB for a large satellite dish antenna.

5.2.2 Gain

Antenna gain is closely related to directivity. The difference is that, for gain, the figure accounts for the efficiency of the antenna so the denominator is referenced not to the radiated power, but to the RF electrical power delivered to the antenna by the transmitter.

$$\text{Antenna gain (dB)} = 10 \times \text{LOG}_{10}$$
$$\text{(Maximum radiated power/RF power delivered to antenna)}$$

If an antenna is 100% efficient, then the directivity and gain are the same. As for directivity, the gain of an antenna can vary from 1.76 dB for a short dipole to as much as 50 dB for a large satellite dish antenna.

5.2.3 Beamwidth

If we are concentrating the radiated power from the antenna in a particular (wanted) direction, then it follows that we will be radiating less power in other directions. Beamwidth is therefore the width of the main beam that radiates in the desired direction. A common measure for beamwidth is the angle where the radiated power drops to half what it was in the direction of the strongest emission; this is called the −3-dB beamwidth.

In the following example, the maximum gain in the desired direction is about 8 dBi. The −3-dB beamwidth is therefore the angle within which the gain remains higher than 5 dBi. If we look at Figure 5.2, we can see that this seems to be roughly +/−10° either side of the desired direction, so we can say that the −3-dB beamwidth in this case is 20°.

Generally, then, the antenna beamwidth and gain have an inverse relationship where the more gain an antenna has, the narrower its −3-dB beamwidth will be.

For omnidirectional antennas, −3-dB beamwidth usually refers to the radiation pattern viewed from the side and can vary from 60° for a simple dipole to 7.5° for a high gain dipole array, whereas, for directional antennas, the −3-dB beamwidth usually refers to the radiation pattern viewed from above and can vary from 60° for a low-gain Yagi antenna to less than 1° for a high gain satellite dish.

5.2.4 Reciprocity

Although not an antenna specification, it is important to understand that, in the case of most antennas, all the transmission properties of an antenna also hold true when the antenna is used to receive. For example, the gain will be the same, no matter whether the antenna is used for transmitting or receiving. This property of antenna is called reciprocity.

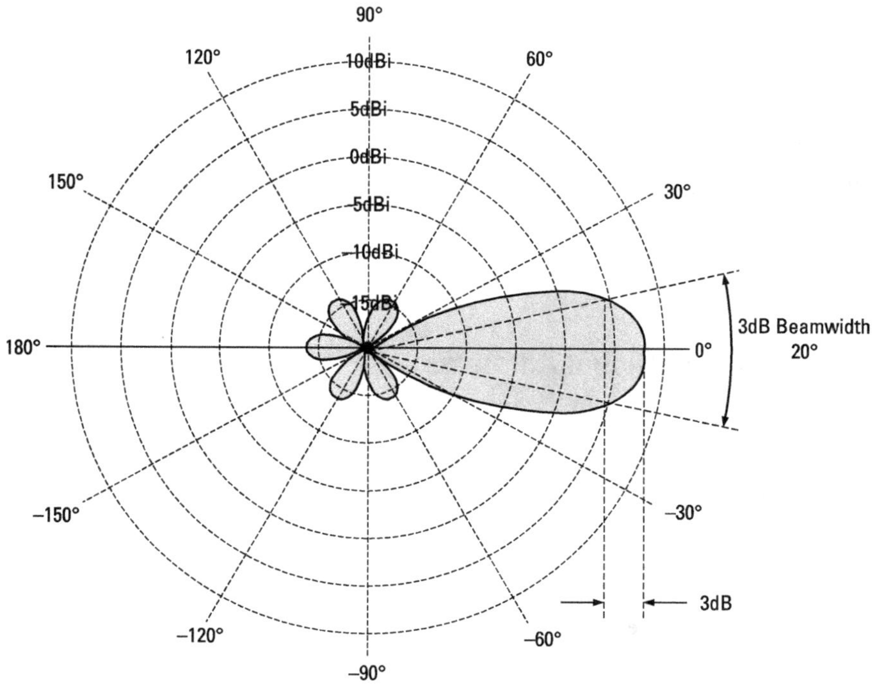

Figure 5.2 Polar plot of a directional antenna viewed from above, showing an antenna with –3-dB beamwidth.

5.2.5 E-Plane

The E-plane is defined as the plane containing the electric field vector. The electric field or E-plane determines the polarization or orientation of the radio wave. For a vertically polarized antenna, the E-plane usually coincides with the vertical/elevation plane. Put more simply, the antenna E-plane polar plot shows the radiation pattern when viewing a vertically polarized antenna from the side.

5.2.6 H-Plane

The H-plane is defined as the plane containing the magnetic field vector and the direction of maximum radiation. The magnetic field or H-plane lies at a right angle to the E-plane. For a vertically polarized antenna, the H-plane usually coincides with the horizontal/azimuth plane. Again, to make it clearer, the antenna H-plane polar plot shows the radiation pattern when viewing a vertically polarized antenna from above.

5.2.7 Azimuth

This is the orientation of the antenna on the tower relative to the magnetic north. It is usually the direction in which we want the strongest emission to go from the site.

5.3 Omnidirectional Antennas

When oriented vertically relative to the surface of the Earth, omnidirectional antennas radiate or intercept RF electromagnetic fields equally well in all horizontal directions. The power radiated drops off with elevation angle so little radio energy is aimed into the sky or down toward the Earth and therefore wasted.

As system cost is proportional to the number of radio repeater sites required to cover the desired area, many of the sites will be at geographical high points. At such sites, omnidirectional antennas are most commonly used because we want coverage in all directions from the site. The following describes some common types of omnidirectional antennas that are used at radio sites.

5.3.1 Half-Wave Dipole

The most basic antenna is the half-wave dipole. As shown in Figure 5.3, it simply consists of two straight conductors, each one quarter of a wavelength

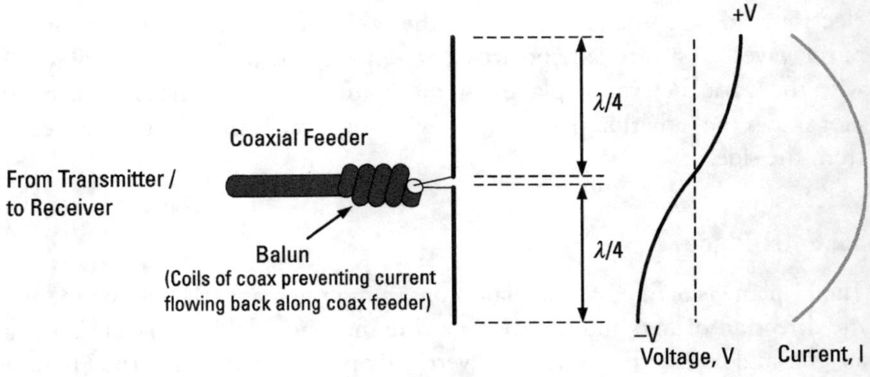

Figure 5.3 Half-wave dipole.

($\lambda/4$) long at the frequency of interest. The overall length then is a half-wavelength ($\lambda/2$).

The RF voltage and current along the half-wave dipole is also shown. The current is 0 at both ends (because the current does not have anywhere to flow), while the voltage is 0 in the middle, rising to a maximum at both ends but in opposite polarity.

Figure 5.4 shows that if viewed from above (H-plane), the half-wave dipole radiates equally in all directions. However, if viewed from the side (E-plane), the radiation is more in front of and behind the dipole, with very little if any radiation directly above and below.

To give a perhaps more real perspective of the radiation pattern, if viewed in 3D, it would look like a doughnut.

On the E-plane plot, the areas of radiation in front of and behind the antenna area are called lobes. As can be seen, the gain of the antenna is 2.15 dB relative to that of an isotropic radiator, which we write as 2.15 dBi.

It can be confusing to visualize how a passive structure like an antenna can have gain. The answer is that the energy is focused; in the case of an omnidirectional antenna, it is focused out to the sides of the antenna as opposed to above and below. Another way to look at it is that the volume of the doughnut-shaped radiation pattern of a half-wave dipole would be the same as that of the spherical radiation pattern of the isotropic radiator.

As the most basic type of antenna that can be practically manufactured, the half-wave dipole is used as a basis against which the performance of other antenna designs can be measured. In other words, because it is difficult to make an isotropic radiator, it is far easier to quantify the performance of an antenna against that of a half-wave dipole instead. The gain of antennas therefore is often quoted in dBd, which is the gain relative to that of the half-wave dipole. Hence:

$$G\ (\text{dBi}) = G\ (\text{dBd}) + 2.15 \tag{5.1}$$

and

$$G\ (\text{dBd}) = G\ (\text{dBi}) - 2.15 \tag{5.2}$$

5.3.2 Half-Wave Folded Dipole

The folded dipole consists of a basic dipole with an added conductor connecting the two ends together to make a complete loop, as shown in Figure 5.5.

Folded dipoles are ideally suited to mounting on the side of towers as

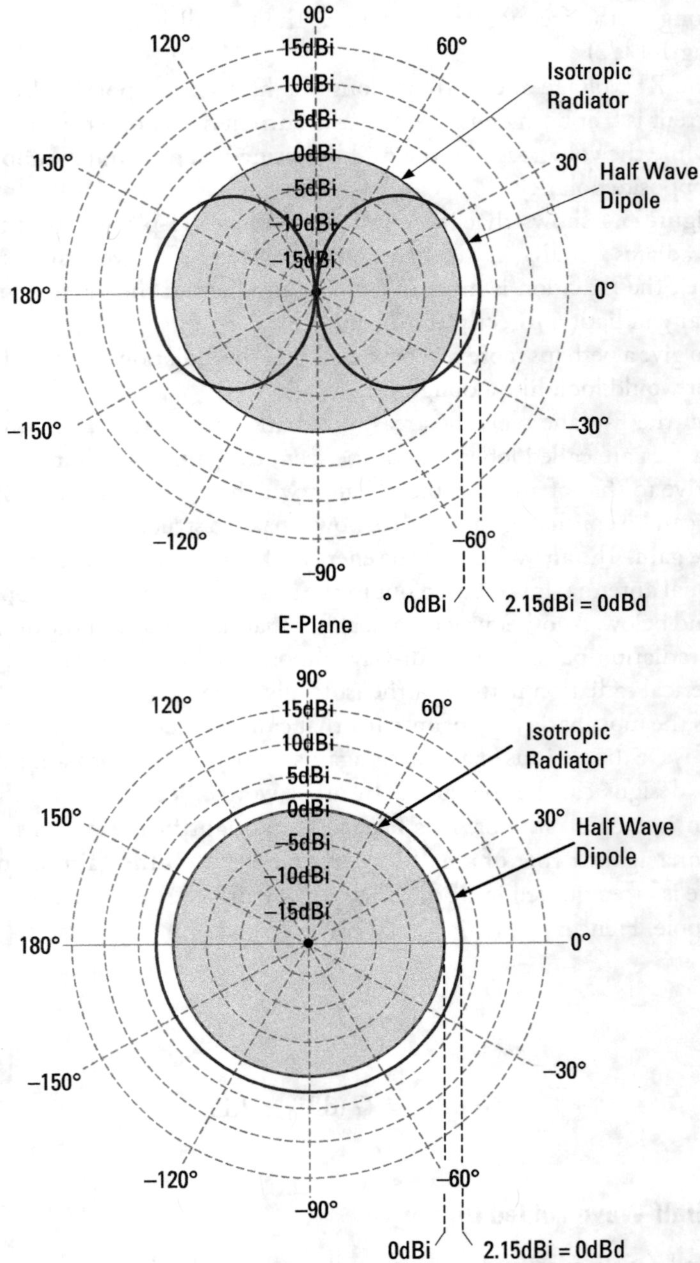

Figure 5.4 Half-wave dipole E and H-plane radiation patterns.

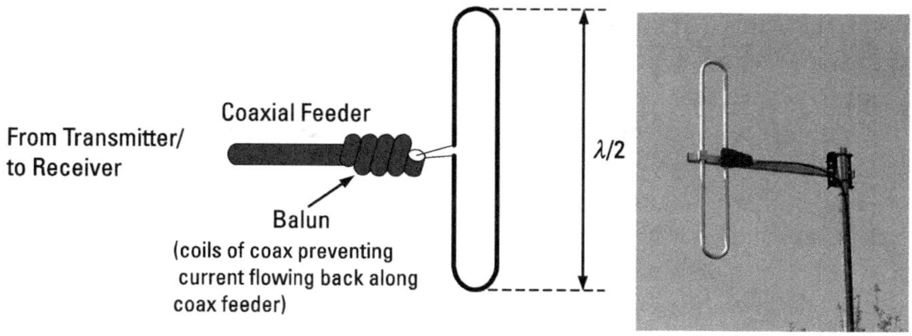

Figure 5.5 Folded dipole.

they are mechanically more rigid than a standard half-wave dipole. They have a higher feed impedance too, so they tend to have a wider bandwidth. The gain and polar plots for a folded dipole are roughly the same as for the standard half-wave dipole.

Half-wave and (more commonly) folded dipoles are a good choice on radio sites where we want equal coverage to all of the surrounding area, but the regulatory limits for radiated power are low (in other words, there is no point using a higher gain antenna to boost the radiated power).

5.3.3 Quarter-Wave Monopole

A quarter-wave ($\lambda/4$) monopole shown in Figure 5.6 is effectively the top section of a half-wave dipole. The monopole antenna was invented in 1895 by radio pioneer Guglielmo Marconi, who discovered if he attached one terminal of his transmitter to a long wire suspended in the air and the other terminal to the Earth, he could transmit for longer distances.

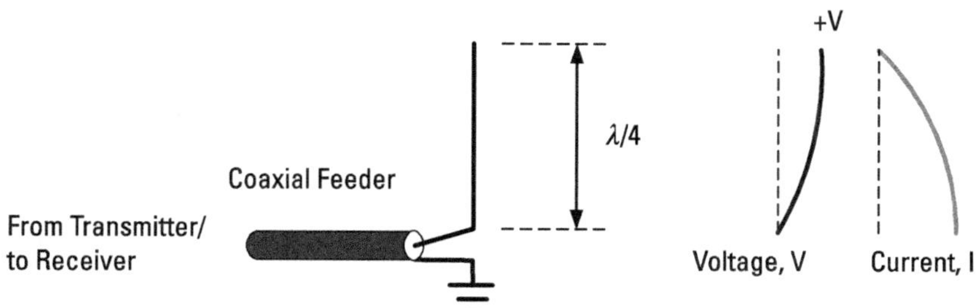

Figure 5.6 Quarter-wave ($\lambda/4$) monopole.

The RF voltage is half that across a half-wave dipole, yet the current is the same. As impedance = V/I, that means the impedance of the $\lambda/4$ monopole is half that of a $\lambda/2$ dipole, or 36 ohms.

If the ground plane below the $\lambda/4$ monopole is large enough, then the radiation pattern is similar to the top half of a dipole pattern (the top half of the doughnut) as shown in Figure 5.7, with its maximum radiation in the horizontal direction, perpendicular to the antenna.

However, because it radiates only into the space above the ground plane, or half the space of a dipole antenna, a $\lambda/4$ monopole antenna will have a gain of 5.15 dBi, twice that of a simple $\lambda/2$ dipole antenna. Monopole antennas are half the size of their dipole counterparts and hence are attractive for real estate savings.

The performance of quarter-wave monopoles is very dependent on the quality of the Earth connection. To exhibit the full 5.15-dBi gain, the ground plane must be perfect and infinite; anything less, the effect is to reduce the gain (sometimes drastically) and raise the direction of highest radiation upwards away from the ground as shown in Figure 5.8. Both are wasteful and result in lost coverage.

5.3.4 Dipole Arrays

At many radio sites, higher gain antennas are required when we are trying to achieve omnidirectional coverage over a wide area. A common way of achieving this is to stack half-wave ($\lambda/2$) folded dipoles at suitable spacings one above the other. Such an antenna is known as a dipole array and examples of these are shown in Figure 5.9.

If all the individual half-wave dipole elements are fed simultaneously, the effect is to increase the gain but decrease the vertical (E-plane) beamwidth. Basically, every time the number of elements is doubled, the gain increases by 3 dB, but the vertical E-plane beamwidth halves as noted in Table 5.1.

Figure 5.10 illustrates this trade-off between gain and vertical (E-plane) beamwidth.

As an analogy, as the number of dipoles in the array is increased, think of the doughnut radiation pattern of Figure 5.4 as being progressively squashed from above and below.

Another advantage of dipole array antennas is that the individual half-wave dipole elements can be pointed in different directions to bring subtle but useful changes in the radiation pattern. Three common examples of this are binary, elliptical, and offset dipole arrays.

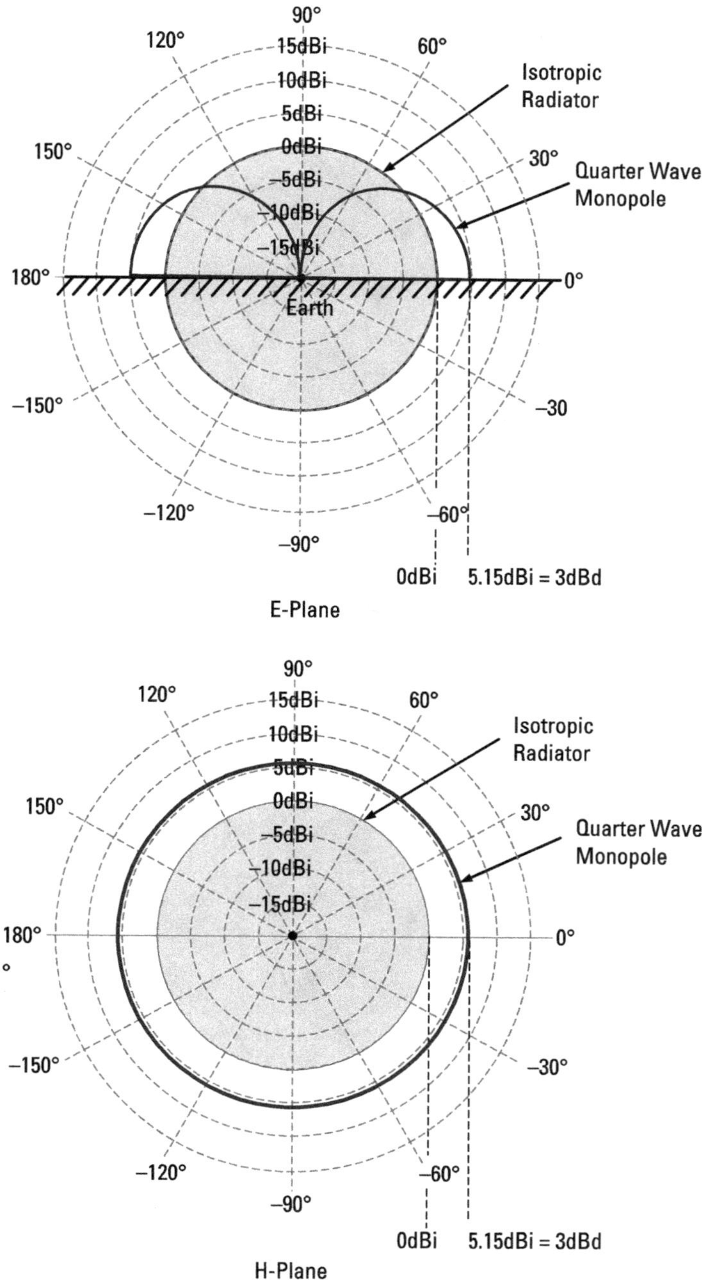

Figure 5.7 Quarter-wave monopole E and H-plane radiation patterns.

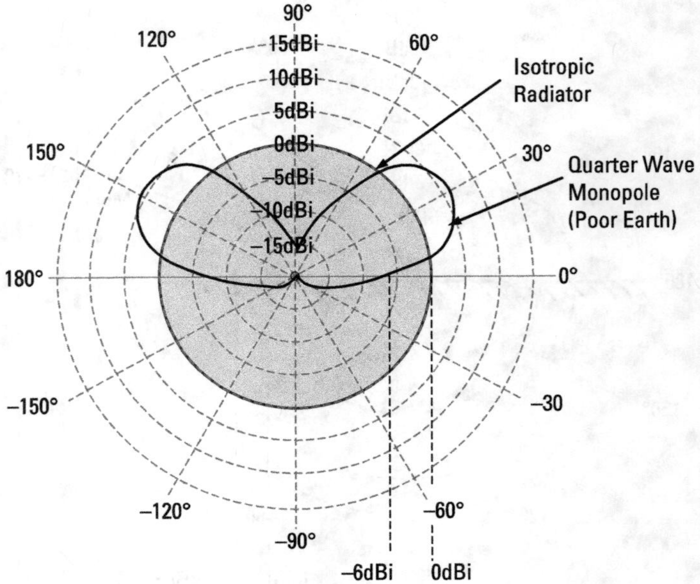

E-Plane

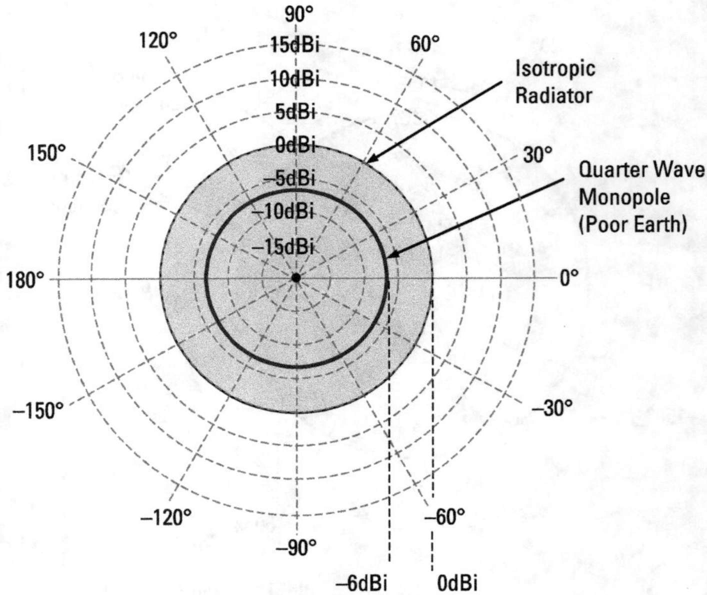

H-Plane

Figure 5.8 Quarter-wave monopole E and H-plane radiation patterns above a poor Earth.

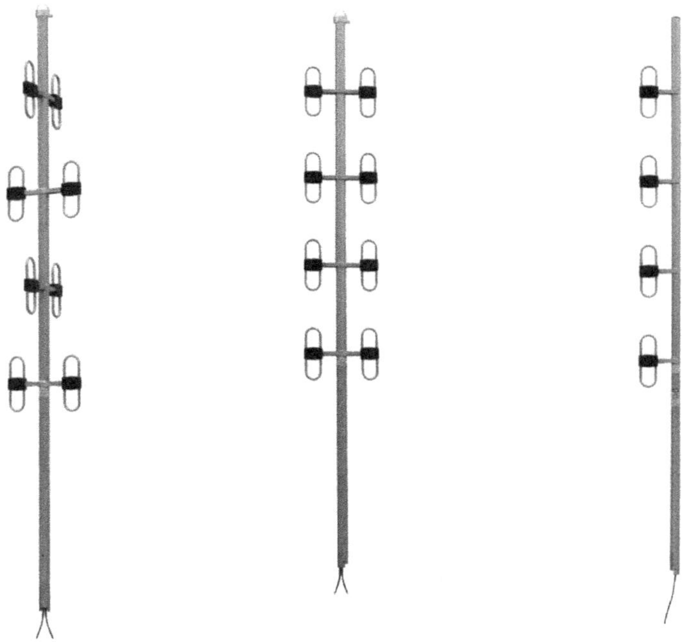

Figure 5.9 Examples of dipole array antennas.

5.3.5 Binary Dipole Array

In a binary dipole array, an equal number of elements point in each direction as shown in Figure 5.11, resulting in coverage that is truly omnidirectional.

For a 4-element binary dipole array like this, the radiation patterns would be as shown in Figure 5.10. Binary arrays are ideal for use on high sites where we want to provide equal coverage to all the lower surrounding area.

Table 5.1
Dipole Array Gain and Vertical Beamwidth

Number of $\lambda/2$ Dipole Elements	Typical Gain	Typical E-Plane (Vertical) Beamwidth
1	0 dBd (2.15 dBi)	60°
2	3 dBd (5.15 dBi)	30°
4	6 dBd (8.15 dBi)	15°
8	9 dBd (11.15 dBi)	7.5°

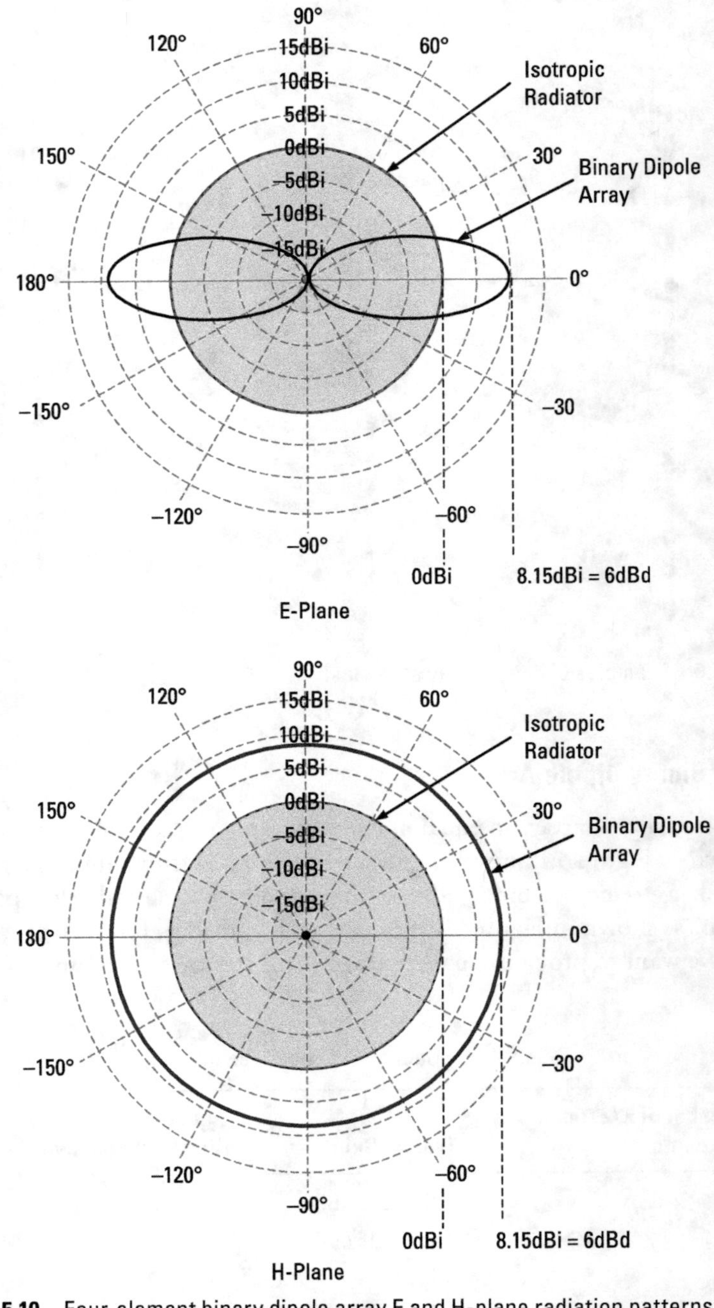

Figure 5.10 Four-element binary dipole array E and H-plane radiation patterns.

Figure 5.11 Binary dipole array antenna.

5.3.6 Elliptical Dipole Array

In an elliptical dipole array, the elements only point in two directions. This results in the elliptical radiation pattern of Figure 5.12, with typically 2-dB more gain than a binary array in the direction in which the dipoles are pointing and slightly less in other directions. They are ideal on sites where a more corridor radiation pattern is required, between two ranges of hills for example.

5.3.7 Offset Dipole Array

In an offset dipole array, all the elements point in the same direction. This results in the offset radiation pattern of Figure 5.13, with typically 3-dB more gain than a binary array in the direction in which all the dipoles are pointing and slightly less in all other directions. Offset arrays are particularly useful on sites that are at one end of the overall area to be covered.

However, as with everything in the field of RF engineering, when something starts to sound too good to be true, then it is too good to be true. There has to be a disadvantage and, in the case of dipole array antennas, it is their vertical size as outlined in Table 5.2.

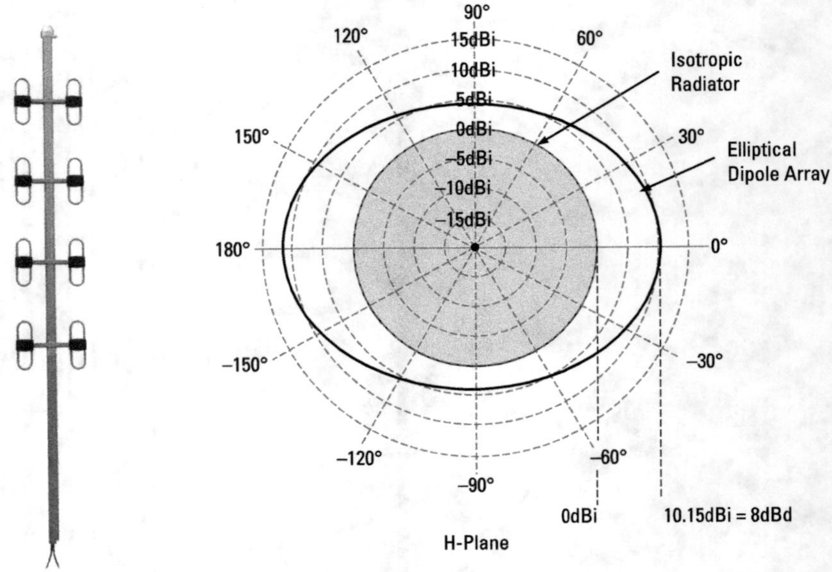

Figure 5.12 Elliptical dipole array antenna and H-plane radiation pattern.

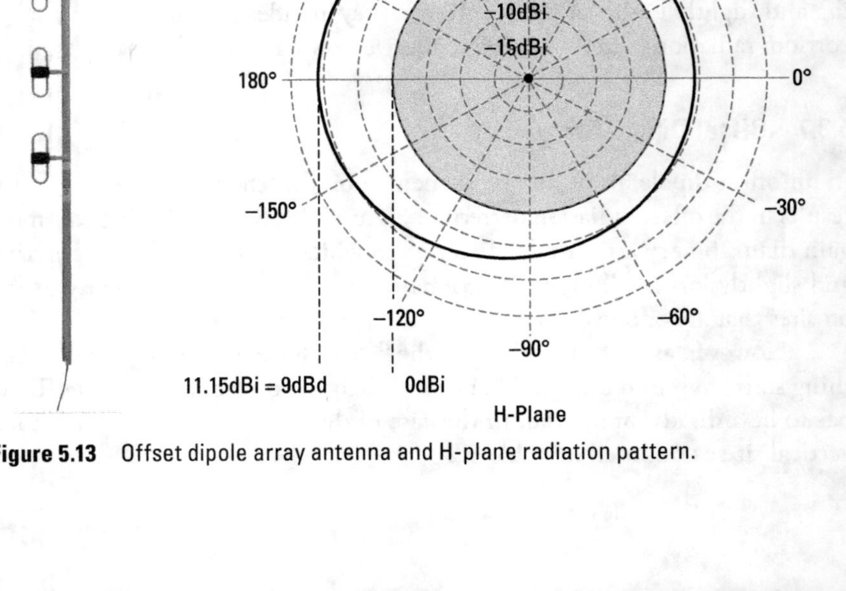

Figure 5.13 Offset dipole array antenna and H-plane radiation pattern.

Table 5.2
Dipole Array Gain and Vertical Beamwidth

LMR Band	Typical 4-Element Dipole Array Height	Typical 8-Element Dipole Array Height
VHF	20 ft (6m)	NA (too high)
UHF	9 ft (3m)	17 ft (5.5m)
800 MHz	5 ft (1.5m)	9 ft (3m)

Another disadvantage of dipole array antennas worth mentioning is, due to their surface area, they are not a good choice on radio sites prone to icing in winter. The weight of ice that can form over the surface of dipole arrays can break the antenna or, worse still, bring it crashing down from the tower.

5.3.8 Colinear Antennas

In the same way that $\lambda/2$ folded dipoles are stacked one above the other to form high gain dipole arrays, $\lambda/4$ monopoles can be similarly configured to form what is known as a colinear antenna.

Again, as with dipole arrays, every time the number of elements is doubled, the gain increases by 3 dB, but the vertical E-plane beamwidth halves. The individual $\lambda/4$ monopoles are housed within a fiberglass tube (known as a radome) as shown in Figure 5.14, meaning the surface area is much less than for dipole arrays. Colinear antennas are therefore a much better choice on radio sites where icing is a problem.

The typical lengths, gains, and vertical beamwidths are the same as those given for dipole arrays in Tables 5.1 and 5.2, and the E and H-plane radiation patterns would be the same as those shown for the binary dipole array in Figure 5.10.

5.3.9 Dual Antennas

Dual antennas usually consist of two colinear antennas or dipole arrays stacked one above the other with a separate feed for each antenna. They are useful at times when there are not enough mounts or enough space left on the tower to accommodate separate Tx and Rx antennas. Dual antennas usually contain two vertically separated antennas for the same band. As the antennas are vertically separated, dual antennas tend to be long, often over 20 ft high.

The isolation between the two antennas is in the order of 30 dB at UHF. If one antenna is used for Tx and the other for Rx, beware, as this is less than half the isolation that could be achieved by mounting physically separate Tx and Rx antennas at different vertical locations on the tower. This shortfall must be made up with extra filtering in the transmitter combining and receiver multicoupling systems at the expense of increased size, cost, and insertion loss. The isolation of a practical dual antenna at VHF would be even less, making dual antennas impractical in this band.

In the antenna shown in Figure 5.15, the top two elements form one binary array, and the bottom two elements are a totally separate binary array. Each has its own feed, hence the two coaxial cables at the antenna base. The gain of both binary arrays in this antenna would be 5.15 dBi (3 dBd).

Figure 5.14 Colinear antenna.

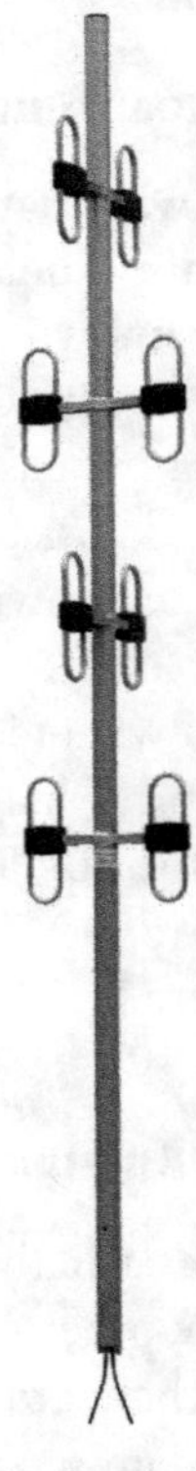

Figure 5.15 Dual two-element binary array antenna.

5.4 Directional Antennas

Compared to an omnidirectional antenna, directional antennas are intended to radiate or intercept RF energy in specific directions only, allowing increased performance and reduced interference from unwanted sources.

Directional antennas are used at radio sites where we only want coverage in some directions from the site, perhaps because the terrain behind the site is shadowed by a range of mountains that are much higher than the radio site, or for point-to-point communications. The following describes some common types of directional antennas that are used at radio sites.

5.4.1 Yagi-Uda

The Yagi antenna was invented by Shintaro Uda in Japan, but the work was presented for the first time in English by Hidetsugi Yagi in 1926, so he tends to get most of the credit. It consists of one radiating element (normally a half-wave dipole or folded dipole), with a reflecting element placed behind the radiator and several directing elements placed in front, as shown by Figure 5.16.

Any signal radiated behind the fed dipole is reflected forwards by the reflector, and the signal radiated forwards is directed ahead by the directors. Hence, the Yagi-Uda antenna is highly directional and radiates little power behind it as illustrated by the polar diagrams of Figure 5.17. Gains of around

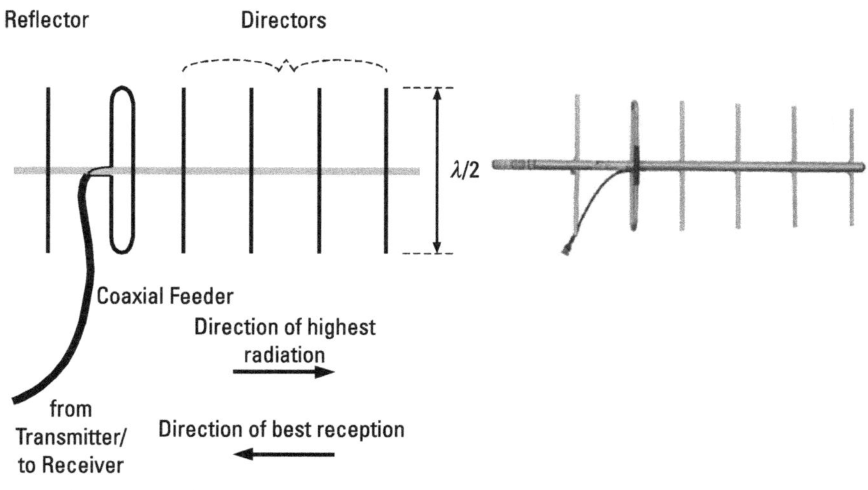

Figure 5.16 Yagi-Uda antenna.

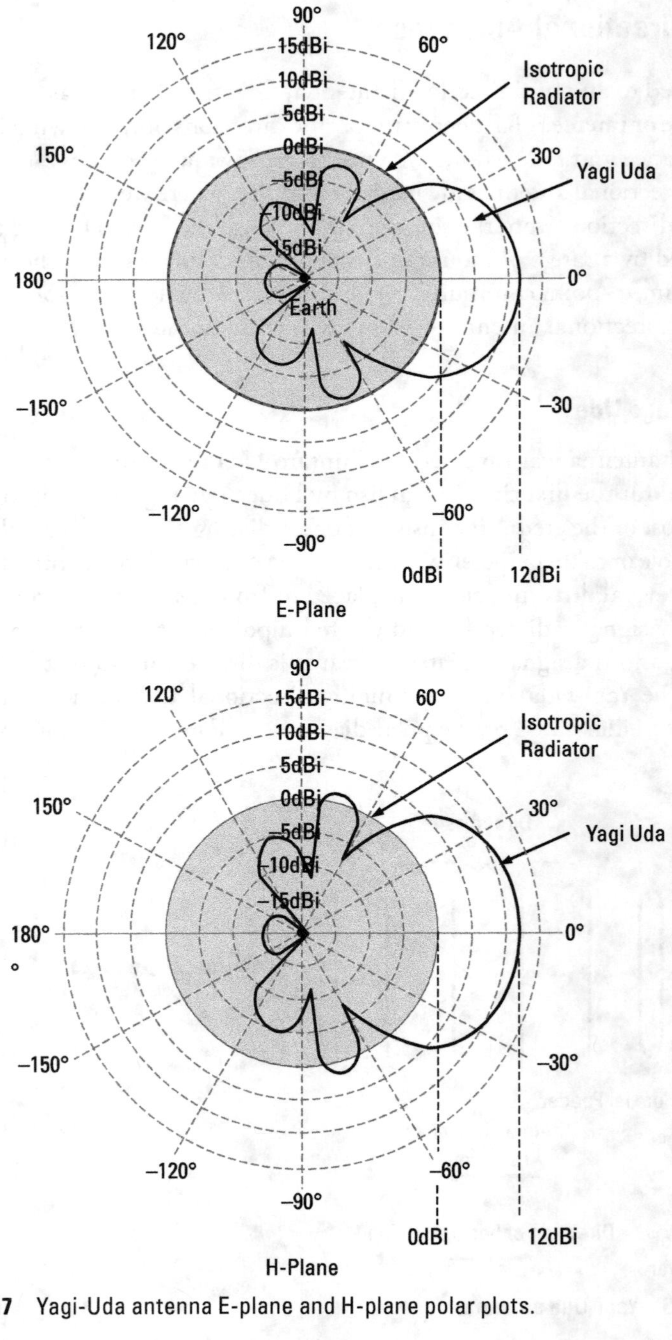

Figure 5.17 Yagi-Uda antenna E-plane and H-plane polar plots.

10 dBi are easily achievable, although the bandwidth is typically quite small. The separation between the elements is around a quarter of a wavelength ($\lambda/4$).

The more directors the antenna has, the more directional it becomes (i.e., the main lobe becomes narrower but the gain in that direction increases). The length of the directors can also be varied to increase the bandwidth over which the antenna will operate efficiently, but at the cost of a little gain.

5.4.2 Panel Antennas

Directional panel antennas are often used at UHF and 800 MHz. These are formed from an array of physical or printed dipoles or monopoles, housed in a fiberglass radome as shown in Figure 5.18. A typical UHF or 800-MHz panel antenna would be about 4 ft (1.25m) high. Horizontal beamwidths of 120° are typical as illustrated in Figure 5.19, so it is common to see three panel antennas orientated on a tower at 0°, 120°, and 240° to give all-around coverage. Such an arrangement is known as a sectorized antenna, and you commonly see them on cellular radio sites.

Mechanical ruggedness is a major advantage of panel antennas. Their short and wide mechanical dimensions mean that they stand up to high wind

Figure 5.18 Panel antenna.

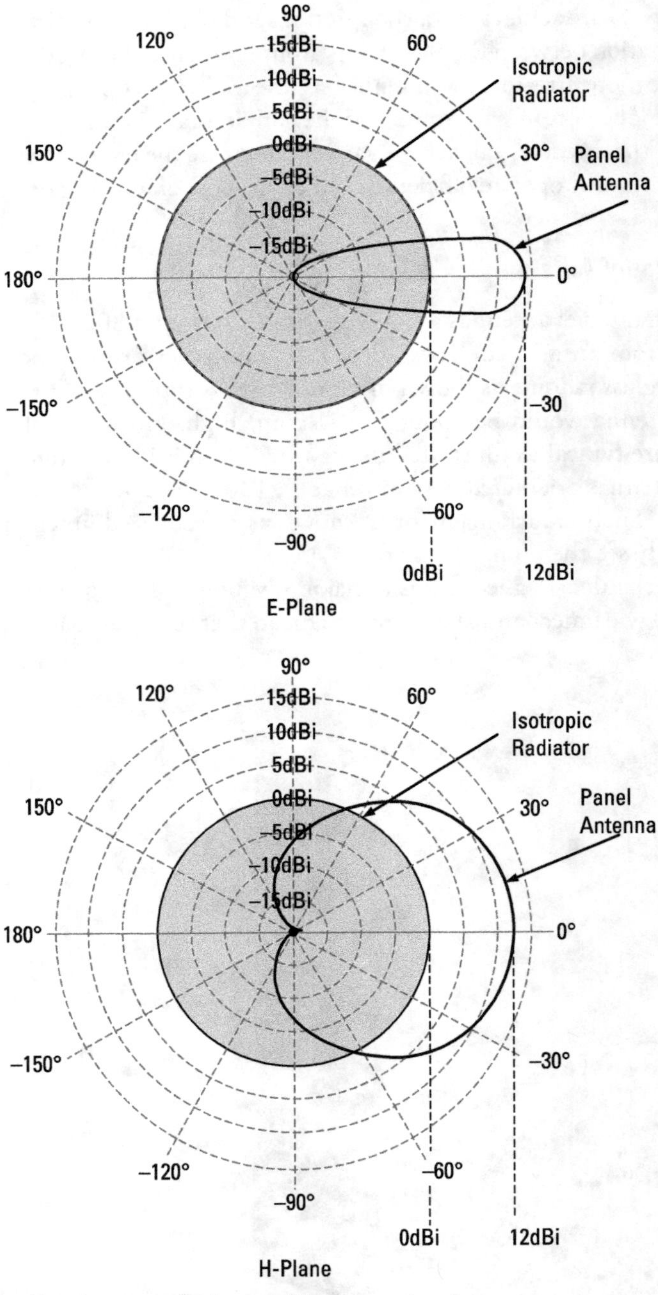

Figure 5.19 Panel antenna E-plane and H-plane polar plots.

or high icing environments when dipole, monopole, or colinear antennas would be prone to snapping.

5.5 Down-Tilt

Before we leave the types of antennas that we might use on radio sites, one last thing to remember is down-tilt. As most antennas are mounted at radio sites on top of hills, and most of the users are on the surrounding lower areas, it is sensible to tilt the antenna downwards a few degrees from vertical as shown in Figure 5.20 so the maximum radiation is propagated in the direction of the users. This can be achieved mechanically, by physically tilting the antenna slightly on the tower, or electrically, by phasing the elements of the antenna accordingly such that the entire radiation pattern is focused slightly downwards.

5.6 Calculating the Isolation Between Antennas on a Radio Tower

Before giving some best-practice design guidelines for placing new antennas on radio towers, it is necessary to understand how to calculate the isolation between neighboring transmit and receive antennas on a radio tower.

To minimize interference between transmitters and receivers at repeater sites to the point where no coverage will be lost, experience shows that there

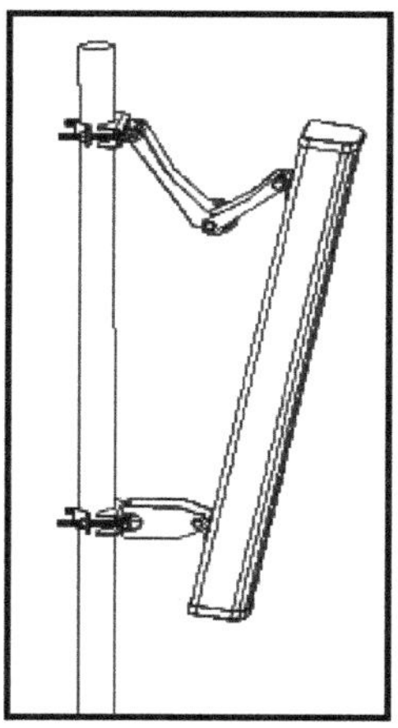

Figure 5.20 Mechanically down-tilted panel antenna.

needs to be around 90-dB total isolation between the site transmitter outputs and the receiver inputs. This total isolation is made up from the sum of the isolation between the associated transmit and receive antennas and the rejection in the transmitter combiner and receiver multicoupler filtering. So, knowing how to calculate antenna isolation is important because it dictates how much rejection we will need in the transmitter combiner and receiver multicoupler filtering.

A key point to understand here is that isolation between the transmit and receive antennas costs nothing, so if 60 dB of the total isolation requirement can be achieved this way, it means only another 30 dB has to be achieved in each of the transmitter combining and receiver multicoupling filter systems, which reduces both the cost and size of those systems, and the losses incurred. Put another way, if we are placing a new receive antenna on the tower, if we can place it so there is 60-dB isolation between it and any transmit antenna, we have achieved most of our interference protection already.

There are three ways in which antennas can be separated on a tower; they can be:

- Horizontally separated;
- Vertically separated;
- Slant separated (separated in both vertical and horizontal planes).

The following is referenced from the ITU-R M.2244 standard.

5.6.1 Isolation Between Horizontally Separated Antennas

The isolation between the horizontally separated omnidirectional antennas shown in Figure 5.21 is calculated using Friis equation for the free-space path loss with the addition of the Tx and Rx antenna gains.

$$I_H = 22 + 20 \times \mathrm{LOG}_{10}\,(d_h\,/\lambda) - (G_{\mathrm{TX}} + G_{\mathrm{RX}}) \tag{5.3}$$

where

I_H = horizontal isolation (dB);

d_h = horizontal separation (m);

λ = wavelength (m);

G_{TX} = gain of Tx antenna (dBi);

G_{RX} = gain of Rx antenna (dBi).

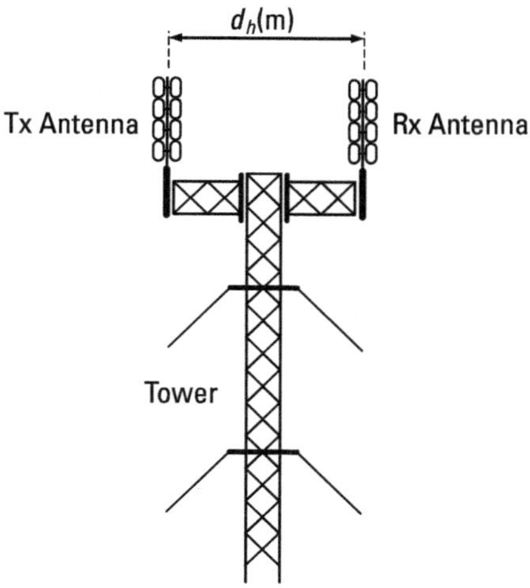

Figure 5.21 Horizontally separated antennas.

For horizontally separated directional antennas, the terms G_{TX} and G_{RX} need to be modified. Basically, it is only the gain of each antenna in the direction of the horizontal separation that is important. Figure 5.22 shows horizontally separated directional antennas set up with different azimuths (the direction the antenna is pointing in relative to the magnetic north pole).

In this case, the horizontal isolation becomes:

$$I_H = 22 + 20 \times \text{LOG}_{10}\,(d_h / \lambda) - (G_{TX(SL)} + G_{RX(SL)}) \qquad (5.4)$$

where

I_H = horizontal isolation (dB);

d_h = horizontal separation (m);

λ = wavelength (m);

$G_{TX(SL)}$ = gain of Tx antenna in the direction of separation (dBi);

$G_{RX(SL)}$ = gain of Rx antenna in the direction of separation (dBi).

So, referring to Figure 5.22, the peak gain of the directional antennas in the direction to which they are pointing is 12 dBi for the transmit antenna (the polar diagram on the left) and 13 dBi for the receive antenna. If these antennas pointed straight at each other, these are the gain figures that we would use, but they do not. Instead, in the direction of the horizontal separation

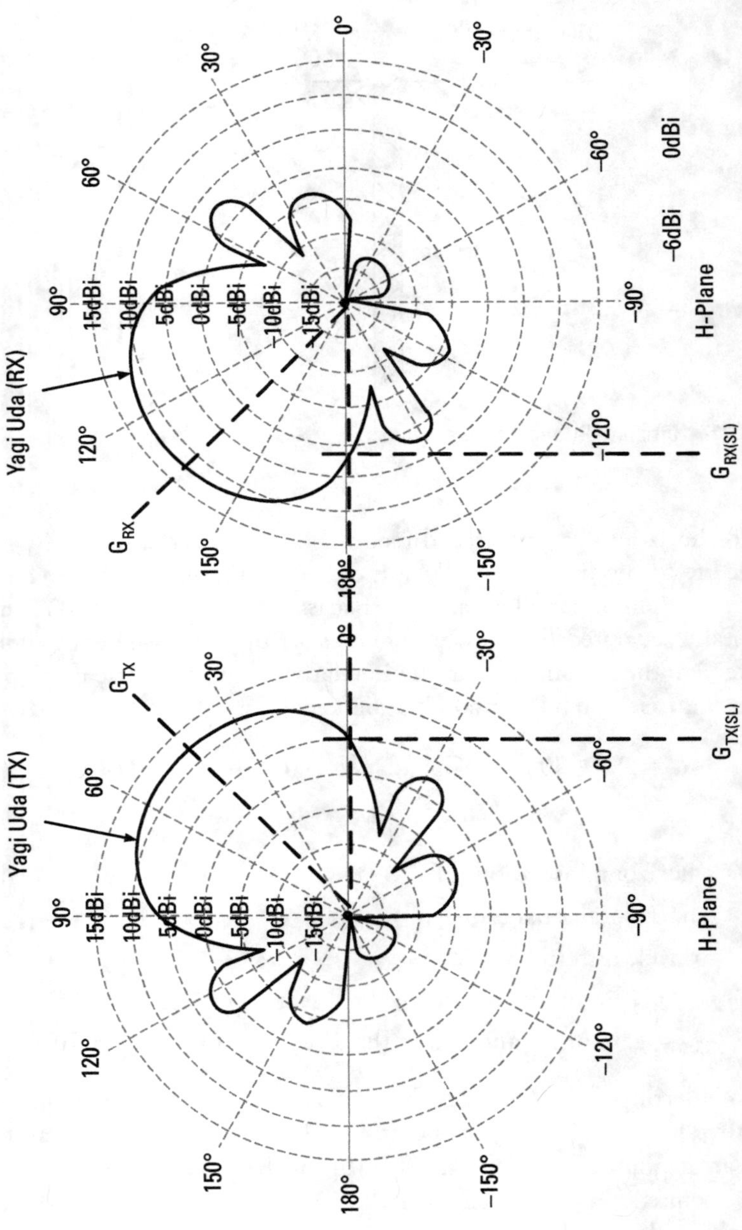

Figure 5.22 Horizontally separated directional antennas.

between them, the gain of the transmit antenna is around 5 dBi and that of the receive antenna has reduced to 3 dBi, so these are the figures that we need to use to calculate the isolation between them.

5.6.2 Isolation Between Vertically Separated Antennas

The isolation between the vertically separated omnidirectional or directional antennas shown in Figure 5.23 is calculated as follows:

$$I_V = 28 + 40 \times \text{LOG}_{10}\ (d_v\ /\lambda) \tag{5.5}$$

where

I_V = vertical isolation (dB);
d_v = vertical separation (m);
λ = wavelength (m).

Note that the vertical separation is measured as the distance between the centers of radiation of the respective antennas, not as the distance between the top of the lower antenna and the base of the higher antenna. Note also that the gains of the antennas do not apply as neither of the antennas have gain in the vertical direction (remember that dipoles have a doughnut shaped radiation pattern).

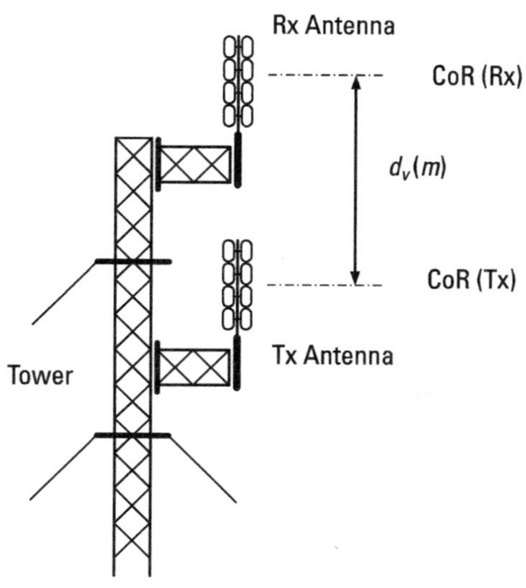

Figure 5.23 Vertically separated antennas.

5.6.3 Isolation Between Slant Separated Antennas

On many occasions, as shown in Figure 5.24, antennas on a tower will be both vertically and horizontally separated. This is known as slant separation and the isolation between antennas separated in this way is:

$$I_S = (I_V - I_H) \times (\alpha/90) + I_H \tag{5.6}$$

where

I_S = slant isolation (dB);

I_H = horizontal isolation (dB);

I_V = vertical isolation (dB);

α = angle between Tx and Rx antennas (degrees).

5.7 Antenna System Topology

For single-channel site RF systems, so long as there is enough frequency separation between the transmit and receive channels, a single duplexed antenna may be used to carry out both the transmit and receive functions as shown in Figure 5.25.

For reasons that will be fully covered in Chapter 7, when the site RF system is multichannel, it is a much better idea to use separate transmit and receive antennas as shown in Figure 5.26.

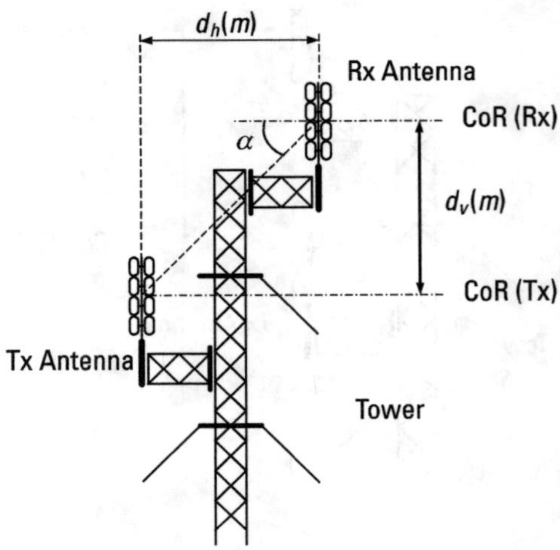

Figure 5.24 Slant separated antennas.

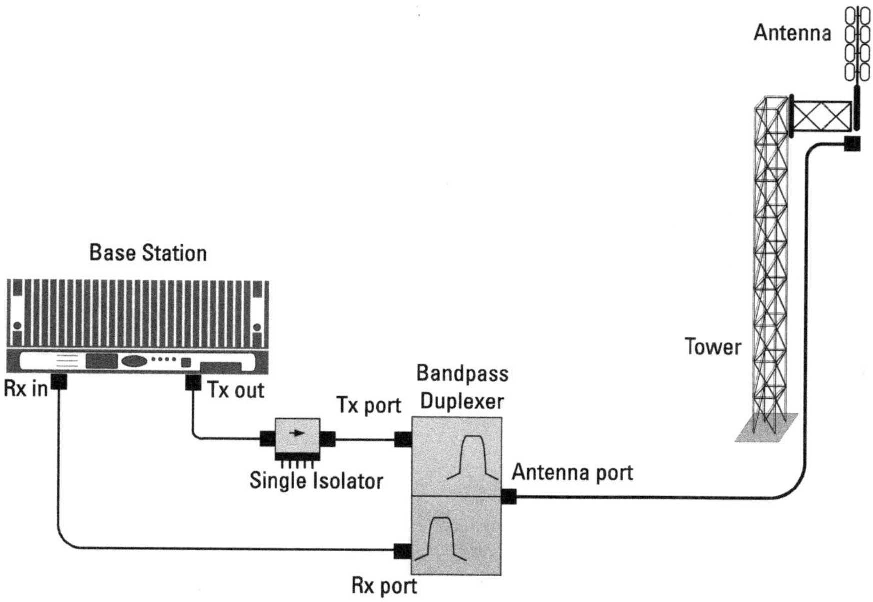

Figure 5.25 Single-channel site RF system with duplexed antenna.

The only times to seriously consider using a single duplexed antenna with a multichannel repeater site RF system are when there simply is not enough space left on the tower for two antennas or because the tower is not rated to take the additional weight. It is not impossible to design interference-free multichannel duplexed systems, but, again, as Chapter 7 explains, it does make interference mitigation much harder.

5.8 Site Antenna System Design Rules

Before concluding this chapter with some best-practice antenna system design rules, it is worth reminding ourselves of the points made in the beginning of this chapter: that the design of a good antenna system is vital and gives the following benefits:

- Maximum coverage;
- Minimum interference potential;
- Minimum transmitter combiner size and cost;
- Minimum receiver multicoupler size and cost.

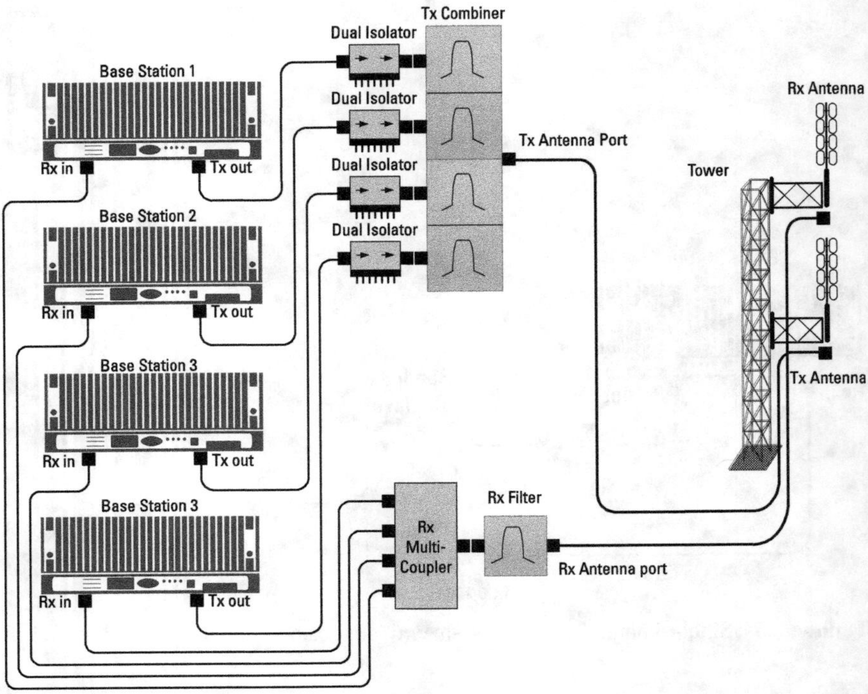

Figure 5.26 Multichannel site RF system with separate transmit and receive antennas.

The following are general best-practice design guidelines that describe how to design optimal antenna systems for deployment on radio towers.

1. Antenna system topology:
   - For single-channel systems, use a single shared transmit/receive antenna.
   - For multichannel systems, use separate transmit and receive antennas. Do not use duplexers for multichannel RF system designs.
2. Place the receive antenna as close as possible to the top of the tower. This maximizes uplink (terminals to base station) coverage.
3. To maximize coverage, minimize interference potential, and optimize transmitter combiner and receiver multicoupler size and cost, design for 60-dB isolation between the transmit and receive antennas:
   - The isolation equations show vertical separation is always preferable.
   - To achieve 60-dB isolation, antennas should be vertically separated on the tower by 12m (40 ft) at VHF, 6m (20 ft) at UHF, and 3m (10 ft) at 800 MHz.

- The transmit antenna should be placed vertically at least 12m below the Rx antenna at VHF, at least 6m below at UHF, and at least 3m below at 800 MHz.

4. No overlap of antennas is permitted.
5. Do not place receive antennas on the tower near existing transmit antennas in the same band. Do not place the receive antenna in a position where there is less than 60-dB isolation between it and any transmit antenna in the same band.
6. Do not place transmit antennas on the tower near existing receive antennas in the same band. Do not place the Tx antenna in a position where there is less than 60-dB isolation between it and any Rx antenna in the same band.
7. Neighboring transmit antennas may be located at the same level on the tower. To prevent the generation of significant IM products (most notably in the isolator outputs), they should be horizontally separated by 4.5m (15 ft) at VHF, 3m (10 ft) at UHF, and 1.5m (5 ft) at 800 MHz.
8. To prevent loss of coverage due to the tower, antennas mounted on the side of the tower should stand off by at least 1.8m (6ft) from the tower leg. Note that when an antenna is mounted at the top of the tower, there is no longer any tower metalwork behind the antenna so it can be mounted directly to the tower leg.
9. Define the tower leg on which each antenna will be deployed and the azimuth (direction relative to magnetic north).
10. All antennas and antenna feeder cables should use 7/16 DIN connectors. As several transmit signals are usually combined onto each transmit antenna at radio repeater sites, the use of 7/16DIN connectors helps to minimize the generation of interference between them caused by passive IM.
11. For antenna feeder cables, as a sensible compromise between cost and RF losses:
    - Use 7/8-inch Heliax cable for antenna feeder cables less than 60m (200 ft) long.
    - Use $1^1/_2$-inch Heliax cable for antenna feeder cables greater that 60m (200 ft) long.

On many occasions, when deploying a new radio system onto a busy site with many existing antennas, the design of an antenna system that meets all these guidelines is simply not possible. All the best positions on the radio towers are already occupied by existing antennas, and all you have left is a choice of all the remaining, less desirable locations. In such cases, you have to make up the shortfall in antenna isolation with additional filtering in both the transmitter combining and receiver multicoupling systems. However, the point is that the closer you can get to meeting these antenna system design rules, the easier the remaining interference mitigation task will be.

Selected Bibliography

dBSpectra, DS4D10F36-U 4 Element Colinear Antenna, https://www.dbspectra.com/products/480-512-mhz-10-dbd-omni-dinf-antenna-ds4d10f36u-d/.

International Telecommunications Union, Report ITU-R M.2244, Isolation Between Antennas of IMT Base Stations in the Land Mobile Service, November 2011.

RFI Technology Solutions, BA80-67 4 Element Binary Array Antenna, https://www.rfi.com.au/BA80-67L-P.

RFI Technology Solutions, BA4040-67 Dual 2 Element Binary Array, https://www.rfi.com.au/BA4040-67-P.

RFI Technology Solutions, BPA3843-65-14 Panel Antenna, https://www.rfi.com.au/BPA3843-65-14.

RFI Technology Solutions, EA80-67 4 Element Elliptical Array Antenna, https://www.rfi.com.au/EA80-67-P.

RFI Technology Solutions, OA40-67 4 Element Offset Array Antenna, https://www.rfi.com.au/OA40-67-P.

RFI Technology Solutions, YB6-99-PIM Yagi Antenna, https://www.rfi.com.au/YB6-99-PIM.

6

Calculating the Interference Threshold

We have discussed the undesirable effects of on-site interference on LMR system coverage, and we have described the various ways that this interference can be generated and deafen a susceptible colocated LMR receiver. The next question then is: to what level must this interference be suppressed at the susceptible site receiver's antenna if we are going to achieve the goal of maximum coverage? To have absolutely no effect on coverage at all, the level of local interference at the receive antenna would need to be zero, which is obviously impossible. Instead, we choose a level of interference that we are willing to accept, with an industry rule of thumb being that which degrades the sensitivity of a susceptible receiver by 1 dB, because this will only have a minimal effect on the distance over which a user can communicate with the site.

This level of interference is called the interference threshold, basically the level below which the sum of any transmitter harmonic, transmitter IM, transmitter sideband noise, receiver blocking, and receiver IM contributions must be suppressed at the base of the receive antenna so as not to result in any more than 1 dB of desensitization of any of the connected receivers. The base of the receive antenna is used as the reference plane because this is where external site noise and interference sources can be added to the thermal noise in the receiver's pass bandwidth and the receive antenna gain accounted for.

The point of this chapter then is to describe how the interference threshold is calculated for a given shared radio repeater site RF system.

Starting at the beginning, as RF system engineers, we live in a world of $C_F / (N + I)$, where:

- C_F = the level of the Rayleigh faded wanted signal at the receiver (dBm);
- N = the noise level in the receiver's bandwidth at the base of the receive antenna (dBm);
- I = the interference level in the receiver's bandwidth at the base of the receive antenna (dBm).

Note that Rayleigh fading is a worst-case multipath propagation model that applies when there is no line of sight between a transmit antenna on a moving vehicle, for example, and the receive antenna at the nearest radio site with which it is communicating. As a result, the signal received by the antenna at the site consists only of multipath components that Rayleigh fading assumes come at equal strength from all directions. At the receive antenna, these multipath components therefore move in and out of phase as the vehicle moves, and so the composite received signal is subject to very short but deep nulls happening at a rate related to the vehicle's velocity.

Our job is to get the Rayleigh faded wanted signal (C_F) far enough above the sum of the noise (N) and interference (I) at the base of the receive antenna to provide at least the minimum acceptable link performance when the signal is demodulated, usually the minimum acceptable Delivered Audio Quality (DAQ) [1] for voice systems or the Message Error Rate (MER) for data systems.

There are several steps involved in calculating the interference threshold, and we will break these down one step at a time to make it clear how the calculation builds up.

6.1 Calculating the Initial Receiver Sensitivity

The calculation starts with working out the susceptible site receiver's thermal noise floor:

$$N_O \text{ (dBm)} = -174 + [10 \times \text{LOG}_{10} (B)] + Rx\ NF \tag{6.1a}$$

where

N_O = thermal noise power referred to the base of the receive antenna (dBm);

B = filter pass bandwidth of the susceptible receiver (Hz);

$Rx\ NF$ = noise figure of the susceptible receiving system (including antenna cabling, filters, multicoupler, and base station receiver) (dB).

As most LMR systems operate in 12.5-kHz channels, practical LMR receivers have a filter pass bandwidth of around 8 kHz. Also, the noise figure of a typical LMR site receiving system is around 8 dB, so using these numbers, we get a thermal noise floor of:

$$\begin{aligned} N_O\ (\text{dBm}) &= -174 + [10 \times \text{LOG}_{10}\ (8 \times 10^3)] + 8 \\ &= -127\ \text{dBm/8 kHz} \end{aligned} \tag{6.1b}$$

The next step is to calculate the best possible receiver sensitivity, which is basically the minimum signal level required at the base of the receive antenna. We do this by adding the relevant $C_F/(N+I)$ threshold to the previously calculated thermal noise floor (N_O); with $C_F/(N+I)$ being the measure of how far the received signal must be above the noise and interference to provide the minimum acceptable level of link performance.

Before going too much further, it is worth pointing out the better the minimum level of performance required, the greater the necessary $C_F/(N+I)$ ratio; in other words, the further above the noise and interference the received signal has to be. Table 6.1 shows some $C_F/(N+I)$ thresholds for common LMR radio system types, taken from the TIA standard TSB-88.1-D [1].

Table 6.1
$C_F/(N+I)$ Thresholds for Common LMR System Types

		$C_F/(N+I)$ Threshold (dB)	
System Type	**Modulation Type**	**DAQ3.0**	**DAQ3.4**
Analog	Narrowband FM	23.0	26.0
Digital FDMA	P25 Phase 1 C4FM	15.2	16.2
Digital TDMA	DMR 4FSK	14.3	15.6
	P25 Phase 2 H-DQPSK	15.2	16.4

Notes: (1) DAQ3.0 corresponds to received audio quality where speech is understandable with slight effort and occasional need for the sender to repeat what they said. (2) DAQ3.4 corresponds to received audio quality that is understandable without repetition but with some noise and distortion present.

It is also just worth touching on a common confusion between $C_F/(N + I)$ and a related parameter that you might also come across, C_S/N. The difference is that $C_F/(N+I)$ is a real-world threshold that includes the Rayleigh fading variation in received signal level that happens when there is no line-of-sight path between the distant user and the receiver at the radio site. The denominator also includes any external noise and interference that adds to the receiver's own thermal noise. C_S/N is the static carrier-to-noise ratio required in a laboratory measurement environment where the receiver under test is directly connected to a signal generator with coaxial cables. C_S/N is always a lower number because the received signal level from the signal generator is constant (no fading), and the only noise and interference present are the thermal noise N_O.

Now, back to the interference threshold calculation. If we assume that our system is a P25 Phase 1 system using C4FM modulation and the customer is asking for a minimum DAQ of 3.0, then the receiver sensitivity becomes:

$$\begin{aligned}\text{Receiver sensitivity (dBm)} &= N_O + [C_F/(N+I)] \\ &= -127 + 15.2 \\ &= -111.8 \text{ dBm}\end{aligned} \tag{6.2}$$

where

Receiver sensitivity = the received signal level required for minimum performance (dBm);

N_O = thermal noise power referred to the base of the receive antenna (dBm);

$C_F/(N+I)$ = the required Rayleigh faded carrier to noise and interference ratio (dB), which, in this case, is 15.2 dB for DAQ3.0 using P25 C4FM modulation (see Table 6.1).

6.2 Adding the Effect of Site Noise

This is the best the receiver sensitivity can be for this site receiving system because the calculation does not yet include any ambient site noise or external interference. To incorporate the site noise into the picture, let us say that a site noise level 10 dB above thermal was measured during the RF survey noise monitoring (see Chapter 2), which, converted to an actual level, is:

$$N_S \text{ (dBm)} = -174 + [10 \times \text{LOG}_{10} (B)] + S \tag{6.3a}$$

where

N_S = site noise referred to the base of the receive antenna (dBm);
B = filter pass bandwidth of the susceptible receiver (Hz);
S = site noise (dB above thermal noise).

So, in this example:

$$\begin{aligned} N_S \text{ (dBm)} &= -174 + [10 \times \text{LOG}_{10} (B)] + S \\ &= -174 + [10 \times \text{LOG}_{10} (8 \times 10^3)] + 10 \\ &= -125 \text{ dBm} \end{aligned} \tag{6.3b}$$

The next step is to work out the total noise (N_T) in the receiver's bandwidth, found by adding the site noise (N_S) to the previously calculated thermal noise (N_O), that is:

$$N_T = 10 \times \text{LOG}_{10} (N_O + N_S) \tag{6.4a}$$

where

N_T = total noise referred to the base of the receive antenna (dBm);
N_S = site noise referred to the base of the receive antenna (mW);
N_O = thermal noise power referred to the base of the receive antenna (mW).

Note the site noise (N_S) and thermal noise (N_O) are in milliwatts and the total noise floor (N_T) is in dBm. This is because site noise (N_S) and thermal noise (N_O) are logarithmic quantities when given in dBm and therefore cannot be added directly; they must be converted back to a linear quantity (milliwatts) first as shown below before they can be summed together.

$$\begin{aligned} N_O \text{ (mW)} &= 10^{(N_O \text{ (dBm)}/10)} \\ &= 10^{(-127/10)} \\ &= 1.99 \times 10^{-13} \text{ mW} \end{aligned} \tag{6.4b}$$

and

$$\begin{aligned} N_S \text{ (mW)} &= 10^{\ (\text{N}_S \text{ (dBm)}/10)} \\ &= 10^{\ (-125/10)} \\ &= 3.16 \times 10^{-13} \text{ mW} \end{aligned} \tag{6.4c}$$

So

$$
\begin{aligned}
N_T &= 10 \times \text{LOG}_{10}\ (N_O + N_S) \\
&= 10 \times \text{LOG}_{10}\ [(1.99 \times 10^{-13}) + (3.16 \times 10^{-13})] \qquad (6.4d) \\
&= -122.88 \text{ dBm}
\end{aligned}
$$

Now we can calculate the receiver sensitivity in the presence of site noise by adding the relevant $C_F / (N + I)$ ratio to the new total noise floor (N_T). If we again assume our system is a P25 Phase 1 system using C4FM modulation and the customer requires a minimum DAQ of 3.0, then:

$$
\begin{aligned}
\text{Receiver sensitivity (dBm)} &= N_T + [C_F / (N + I)] \\
&= -122.88 + 15.2 \qquad (6.5) \\
&= -107.68 \text{ dBm}
\end{aligned}
$$

where

Receiver sensitivity = the received signal level required in the presence of site noise for minimum performance (dBm);

N_T = sum of thermal and site noise referred to the base of the receive antenna (dBm);

$C_F / (N + I)$ = the Rayleigh faded carrier-to-noise and interference ratio from Table 6.1 (dB).

This illustrates the importance of doing the site RF survey and finding out what the site noise floor really is. Before the site noise was added, the receiver sensitivity in the presence of thermal noise was −111.8 dBm. Now, with site noise 10 dB above thermal added, the receiver sensitivity has degraded to −107.68 dBm, a reduction of just over 4 dB, which will cause a noticeable decrease in the distance from which a user in the field can be heard. Exactly how much distance would be lost depends on the terrain, but over an uninterrupted flat landscape, 4 dB of lost sensitivity would equate to kilometers.

6.3 Calculating the 1-dB Interference Threshold

Now we are getting to the business end of the calculation, as the next step is to work out the interference threshold itself, the level of interference that adds to the total noise floor (N_T) and causes 1 dB of receiver desensitization that we decided to tolerate earlier.

So, let's say:

$$N_X = 10 \times \text{LOG}_{10}\,(N_T + I_T) \tag{6.6}$$

where

N_X = total noise and interference floor referred to the base of the receive antenna (dBm);

N_T = total noise floor referred to the base of the receive antenna (mW);

I_T = interference threshold, the total interference level referred to the base of the receive antenna (mW).

Transposing this to find the interference threshold (I_T), the equation becomes:

$$I_T = 10 \times \text{LOG}_{10}\,(N_X - N_T) \tag{6.7}$$

where

I_T = interference threshold, the total interference level referred to the base of the receive antenna (dBm);

N_X = total noise and interference floor referred to the base of the receive antenna (mW);

N_T = total noise floor referred to the base of the receive antenna (mW).

Again, please note the units in the above, as N_X and N_T must be converted to milliwatts first before they can be added or subtracted.

The problem at the moment is that we do not know the values of N_X or I_T. So, let us look at it another way; that is, if the total level of interference (I_T) that we are going to add to the −122.88-dBm total noise floor (N_T) causes the sum (N_X) to increase by 1 dB, we can say that:

$$\begin{aligned} N_X &= N_T + 1\ \text{dB} \\ &= -122.88 + 1 \\ &= -121.88\ \text{dBm} \end{aligned} \tag{6.8}$$

Now that we know the values of N_X and N_T, all we need to do is convert them to milliwatts and we can find the interference threshold, I_T.

$$N_X\,(\text{mW}) = 10^{(N_X\,(\text{dBm})/10)}$$

$$
\begin{aligned}
&= 10^{(-121.88/10)} \qquad (6.9)\\
&= 6.49 \times 10^{-13} \text{ mW}
\end{aligned}
$$

and

$$
\begin{aligned}
N_T\,(\text{mW}) &= 10^{(N_T\,(\text{dBm})/10)}\\
&= 10^{(-122.88/10)} \qquad (6.10)\\
&= 5.15 \times 10^{-13} \text{ mW}
\end{aligned}
$$

Then, by substituting in these values, we find the interference level (I_T):

$$
\begin{aligned}
I_T &= 10 \times \text{LOG}_{10}\,(N_X - N_T)\\
&= 10 \times \text{LOG}_{10}\,[(6.49 \times 10^{-13}) - (5.15 \times 10^{-13})] \qquad (6.11)\\
&= -128.75 \text{ dBm}
\end{aligned}
$$

That is the total level of interference that we can tolerate at the base of the receive antenna if we are only to suffer a further 1 dB of receiver desensitization in addition to that caused by the site noise.

6.4 Accounting for Multiple Interference Mechanisms

The final part of the problem is that there can be multiple interference mechanisms at play. As described in Chapter 4, local interference at a shared radio site can be in the form of:

- Harmonics from other transmitters;
- Transmitter IM, where products generated within a transmitter combiner are radiated at the same frequency to which a nearby receiver is tuned;
- Transmitter sideband noise from colocated transmitters that falls within the susceptible receiver's pass bandwidth;
- Receiver blocking, where a very strong off-channel signal from a nearby transmitter overloads the susceptible receiver;
- Receiver IM produced in the site receiving system because of all the strong interfering signals that appear on the receive antenna.

While it is quite unlikely that all these mechanisms would occur simultaneously at a given shared radio site, it is certainly possible that more than one mechanism is having an influence. For example, there may be no harmonics, or no transmitter or receiver IM products that land on your receive frequency, but there could easily be nearby transmissions that could deafen the receiver by means of sideband noise and blocking. If only one of these mechanisms is occurring at the radio site in question, then the interference per mechanism is the same as the total interference level, I_T. If two of the five mechanisms were present, then the tolerable interference per mechanism is equal to $I_T - 3$ dB and so forth. In other words:

$$\text{Interference per mechanism (dBm)} = I_T - [10 \times \text{LOG}_{10}(n)] \quad (6.12)$$

where

I_T = interference threshold, the total interference level at the base of the Rx antenna (dBm);

n = number of interference mechanisms simultaneously occurring.

Reducing the total interference to this level is the role of the filtering in the transmitter combiner and receiver multicoupler systems, and we will learn how this is done in Chapters 8 and 9.

6.5 Recalculating the Final Receiver Sensitivity

The last part of the calculation then is to recalculate the receiver sensitivity in the presence of thermal noise, site noise, and interference by adding the relevant $C_F/(N+I)$ ratio to the new total noise and interference floor (N_X). As before, if we assume that our system is a P25 Phase 1 system using C4FM modulation and the customer requires a minimum DAQ of 3.0, then:

$$\begin{aligned}\text{Receiver sensitivity (dBm)} &= N_X + [C_F/(N+I)] \\ &= -121.88 + 15.2 \quad (6.13) \\ &= -106.68 \text{ dBm}\end{aligned}$$

where

Receiver sensitivity = the received signal level required in the presence of site noise for minimum performance (dBm);

N_X = total noise and interference floor referred to the base of the receive antenna (dBm);

$C_F/(N+I)$ = the Rayleigh faded carrier to noise and interference ratio from Table 6.1 (dB).

Before the site noise was added, the receiver sensitivity in the presence of thermal noise only was −111.8 dBm. Now, with site noise 10 dB above thermal added, along with the permissible level of interference, the receiver sensitivity has degraded to −106.68 dBm, a reduction of over 5 dB that could potentially decrease in the distance from which a user in the field can contact the site by several kilometers.

Figure 6.1 attempts to illustrate the situation by depicting an ambulance at such a distance from the radio site that the mean received signal level at the

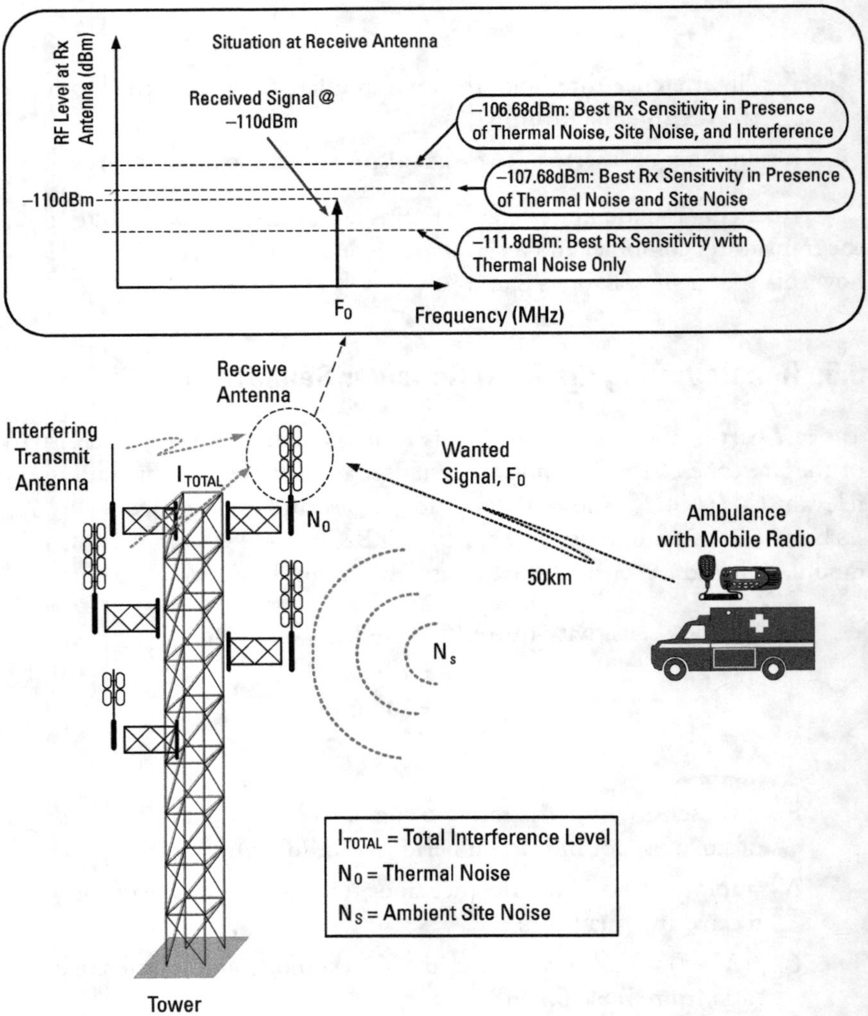

Figure 6.1 Effect of site noise and interference on receiver sensitivity.

site is −110 dBm. In the presence of thermal noise only, the sensitivity of the site receiver is −111.8 dBm, meaning the received signal can be demodulated into intelligible speech of a quality slightly better than DAQ3.0. However, once site noise is added, the associated rise in the site receiver's noise floor means that the lowest signal that can now be detected is −107.68 dBm, some 2 dB greater than our received signal. The effect would be a noticeable decrease in the received audio quality, to the point where frequent repetition of the message will be necessary. With the addition of interference, which we allow to make the sensitivity 1 dB worse, the intelligibility of the received audio decreases further. Effectively then, it is because of noise and interference at the radio site that the signal from the ambulance cannot be received adequately. This shows why it is essential to quantify both the site noise and interference threshold before trying to estimate the coverage.

6.6 Summary

By now, you can see that it takes a lot of separate calculations and even more words to explain how to calculate the interference threshold. So, for those who prefer everything summarized in one place, Figure 6.2 is for you.

In Figure 6.2:

$$N_O \text{ (dBm)} = -174 + [10 \times \text{LOG}_{10} (B)] + Rx\ NF \tag{6.14}$$

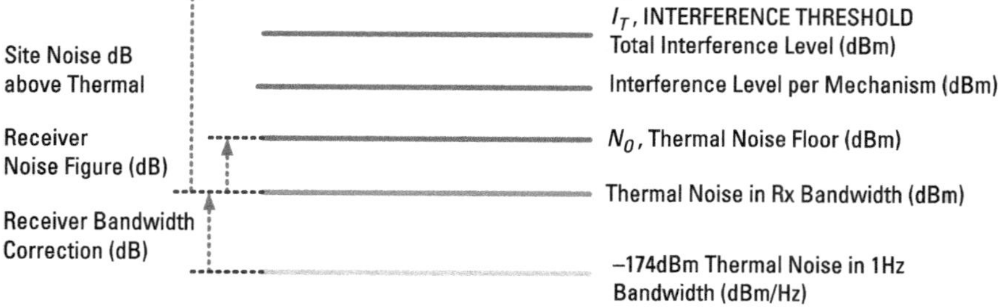

Figure 6.2 Calculation of interference threshold.

where

N_O = thermal noise power referred to the base of the Rx antenna (dBm);

B = receiver equivalent noise bandwidth (Hz);

$Rx\ NF$ = noise figure of the susceptible receiving system (including antenna cabling, filters, multicoupler, and base station receiver) (dB).

$$N_S\ (\text{dBm}) = -174 + [10 \times \text{LOG}_{10}\ (B)] + S \tag{6.15}$$

where

N_S = site noise referred to the base of the receive antenna (dBm);

B = receiver equivalent noise bandwidth (Hz);

S = site noise (dB above KTB).

$$N_T = 10 \times \text{LOG}_{10}\ (N_O + N_S) \tag{6.16}$$

where

N_T = total noise referred to the base of the receive antenna (dBm);

N_S = site noise referred to the base of the Rx antenna (mW);

N_O = thermal noise power referred to the base of the Rx antenna (mW).

$$\begin{aligned} N_X &= N_T\ (\text{dBm}) + 1\ \text{dB} \\ &= 10 \times \text{LOG}_{10}\ (N_T + I_T) \end{aligned} \tag{6.17}$$

where

N_X = total noise and interference floor referred to the base of the receive antenna (dBm);

N_T = total noise floor referred to the base of the receive antenna (mW);

I_T = interference threshold, the total interference level referred to the base of the receive antenna (mW).

$$I_T = 10 \times \text{LOG}_{10}\ (N_X - N_T) \tag{6.18}$$

where

I_T = interference threshold, the total interference level referred to the base of the receive antenna (mW);

N_X = total noise and interference floor referred to the base of the receive antenna (mW);

N_T = total noise floor referred to the base of the receive antenna (mW).

$$\text{Interference per mechanism (dBm)} = I_T - [10 \times \text{LOG}_{10}\ (n)] \quad (6.19)$$

where

I_T = interference threshold, the total interference level referred to the base of the receive antenna (mW);

n = number of interference mechanisms simultaneously occurring.

$$\text{Receiver sensitivity (dBm)} = N_X + [C_F\ /(N + I)] \quad (6.20)$$

where

Receiver sensitivity = the received signal level required in the presence of site noise for minimum performance (dBm);

N_X = total noise and interference floor at the base of the receive antenna (dBm);

$C_F\ /(N + I)$ = the Rayleigh faded carrier to noise and interference ratio (dB).

Reference

[1] Telecommunications Industry Association, Report TSB-88.1-F, Wireless Communications Systems—Performance in Noise and Interference Limited Situations—Part 1: Recommended Methods for Technology Independent Narrowband Performance Modelling, April 2022.

7

Duplexers

Having looked at antennas, it is now time to focus on the other main parts used in repeater site RF systems. That is what the next three chapters are dedicated to, starting here with the subject of duplexers.

In your LMR network, it is quite possible that some sites will only need to be single-channel and the RF system can be based on a duplexer, whereas other sites will be multichannel and require transmitter combiners and receiver multicouplers. There may even be some multichannel sites that only have room for only one more antenna on the tower and we will need a transmitter combiner, a receiver multicoupler, and a duplexer.

To make sense of the rest of this chapter, let us first understand what our task is and is not. The task at hand is not to physically design and make the duplexer, as you can buy high-quality duplexers suitable for deployment on radio repeater sites from several reputable vendors. Instead, the vital task that we have is to choose the right duplexer for each site that needs one. This entails:

- Understanding the various types of duplexers and how they work. This gives us the insight to choose the right type of duplexer for each site.
- Specifying the performance that we need from the duplexer, which might be quite different for each site depending on what strong interfering signals already exist there and where their antennas are located.

Onto the subject of duplexers, in RF engineering, you may come across two similarly named devices: duplexers and diplexers. Defining the difference can get confusing and becomes even more so when you consider that the same two filters in parallel can be used as either a duplexer or a diplexer.

So, let us look at a simple way to define what a duplexer is. The simplest form of RF communication is called simplex, where the users talk directly to one another with radios that use one frequency and can only transmit or receive at any one time. Duplex RF communication is more complicated but much more flexible. Duplex radios use two frequencies per channel, one to transmit and one to receive. There are two types of duplex, called half-duplex where the radio can only transmit or receive at any one time and full-duplex where the radio can do both at once. In the context of LMR, remember that LMR systems are based on terminals (vehicle-mounted mobiles and user handportables) that, rather than communicate directly, do so via fixed repeaters located at geographically high sites which receive the transmission from the sender's terminal and retransmit (or repeat) it to the intended recipient so the distance over which the users can communicate is greatly increased. To operate like this, the user terminals only need to work in half-duplex, but the repeaters must be full-duplex because they can be transmitting (repeating) information from one user at the same time as they are receiving information from someone else.

So, the easiest definition of a duplexer is a device that allows a repeater connected to one antenna to operate in full-duplex mode.

Figure 7.1 shows that a duplexer is basically a three-port passive device consisting of two filters in parallel that allows a transmitter and a receiver operating at different frequencies to share the same antenna. One of the filters passes the transmitted signal to the common antenna and suppresses transmitter noise at all other frequencies, including the receive frequency. The other filter passes the desired signal picked up by the common antenna on to the receiver while rejecting the transmit signal and any unwanted received signals. The base station transmitter is therefore isolated from its receiver because the passbands of the duplexer transmit and receive filters do not overlap.

7.1 Types of Duplexers

The duplexers used in repeater site RF systems are usually constructed in one of two ways. The most common is the milled duplexer where the transmit and receive filters are machined from one solid block of aluminum. Figure 7.2 shows a typical milled duplexer construction; the studs extending through

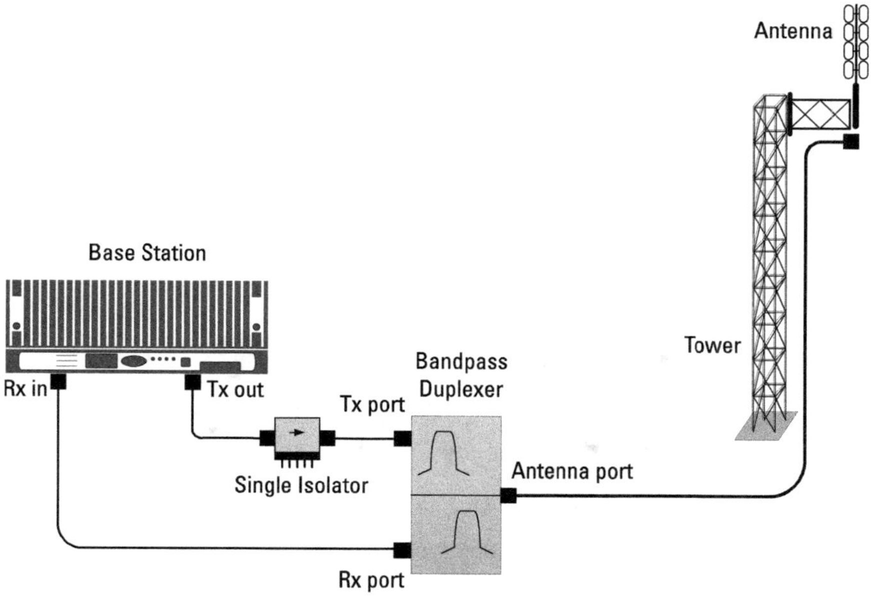

Figure 7.1 Use of a duplexer in a single-channel site RF system.

the lid are tuning rods that fit down inside the machined cavities so the exact frequency response of the filters can be optimized. This kind of duplexer can be used when we have standardized transmit to receive frequency separations of 4.6 MHz at VHF, 9 MHz at UHF, and >30 MHz in the 800/900-MHz LMR bands.

However, there are cases where the transmit to receive frequency separation is much less, a prime example being the VHF systems in North America. This band was originally used for simplex communications, so there was no need to regulate which parts of the band were used for transmit and which parts were used for receive. The consequence, now that band is used for duplex communications too, is that the separation between the transmit and receive frequencies can be extremely narrow, as little as 500 kHz. Milled duplexers cannot achieve the required rejection (around 70 dB) at such narrow offsets, so much larger (about 5 times) cavity filter-based duplexers like the example shown in Figure 7.3 have to be used instead.

As with most filters, duplexers are available in the bandpass, band-reject, bandpass/band-reject, and highpass/lowpass configurations shown in Figure 7.4.

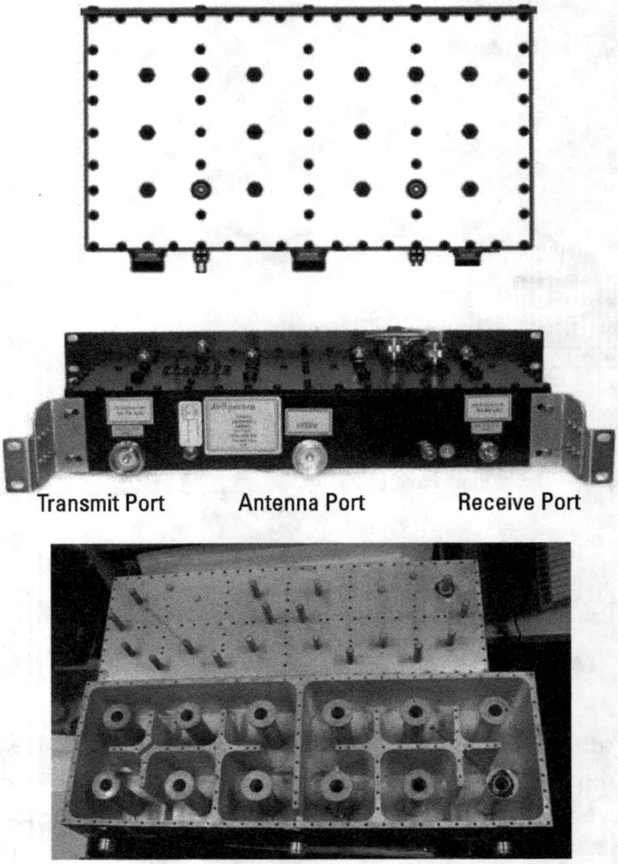

Figure 7.2 Construction of a milled duplexer.

A very important point to which to draw attention here is only use bandpass duplexers on shared radio sites. Specifically, do not be tempted to use band-reject (notch) duplexers. These may typically be smaller and cheaper, but consider what would happen. First, think about the duplexer transmit to antenna path, which has a notch at the receive frequency but passes all other frequencies. What happens is the notch rejects the sideband noise from your connected transmitter at your connected receiver's frequency; however, the transmitter noise at all other frequencies is radiated with very little attenuation and so has the potential to deafen other site receivers with nearby antennas. Similarly, in the antenna to receive path of a band-reject duplexer, there is a notch at the connected transmitter's frequency. This rejects the transmitted signal and prevents it from deafening (blocking) the connected receiver.

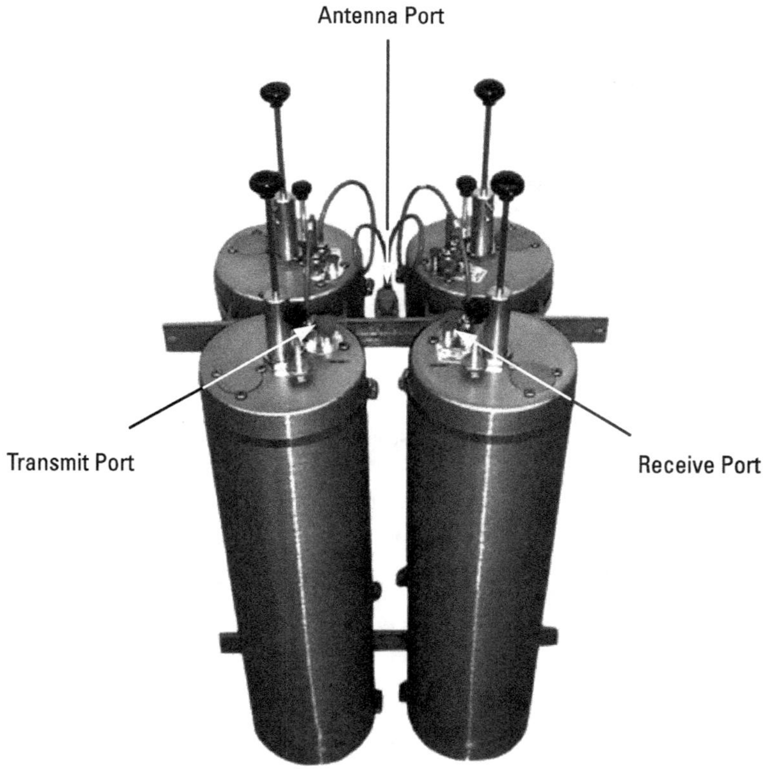

Figure 7.3 Cavity filter-based duplexer for narrow transmit to receive frequency offsets.

However, any high-level signals at other frequencies, picked up from other site transmitters with nearby antennas, are not rejected and can pass to your receiver resulting in a loss of sensitivity due to blocking or IM. The same problems occur with highpass/lowpass and bandpass/band-reject duplexers, albeit to a lesser extent. So, bandpass duplexers are really the only sensible choice for use on shared radio sites.

7.2 Specifying Duplexer Design Requirements

To derive the specification for a bandpass duplexer suitable for deployment on a shared radio site, we need to consider the two filters separately. The filter between the transmit and antenna ports of the filter must pass the transmitted

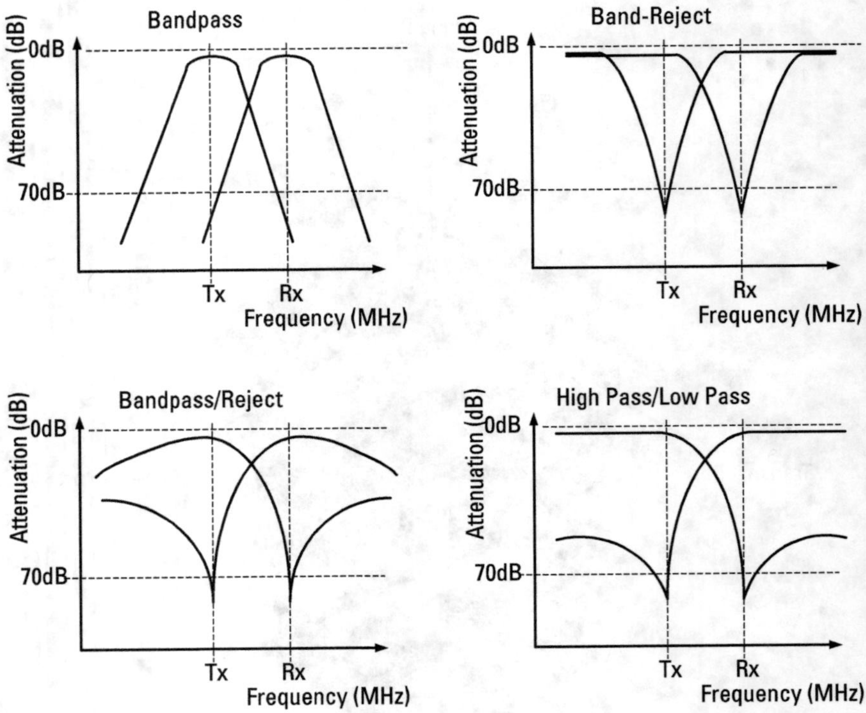

Figure 7.4 Types of duplexers.

signal at minimum loss and suppress the transmitter sideband noise at all other frequencies, especially at the frequency of the receiver sharing the antenna. To calculate the rejection required at the receive frequency in the duplexer transmit to antenna filter, we need to know the transmitter power at the duplexer antenna port, the transmitter sideband noise level at the receive frequency, and also the interference threshold below which the transmitter sideband noise must be suppressed at the receiver input so as not to result in any more than 1 dB of desensitization of the receiver.

As an example, let:

- Transmitter power at duplexer antenna port, P_{TX} = +50 dBm (or 100W);
- Transmitter sideband noise at receive frequency, $Tx\ SBN_O = -150$ dBc/Hz;
- Receiver equivalent noise bandwidth, $B = 8$ kHz;
- Receiver noise figure, $Rx\ NF = 8$ dB;

- Rayleigh faded carrier-to-noise and interference ratio, $C_F\ /(N+I) =$ 15.2 dB: for a P25 Phase 1 system using C4FM modulation where a minimum DAQ of 3.0 is required (see Chapter 6, Table 6.1);
- Static carrier-to-noise ratio for the same delivered audio quality, $C_S\ /N = 8$ dB.

7.2.1 Transmit to Antenna Filter

To calculate the rejection required in the transmit to antenna filter, we start at the transmit end by calculating the transmitter sideband noise present at the antenna port on the receive frequency:

$$\begin{aligned}&Tx\ SBN\ (\text{dBm/8 kHz})=\\&Tx\ SBN_O\ (\text{dBc/Hz}) + P_{\text{TX}} + [10 \times \text{LOG}_{10}\ (B)]\end{aligned} \quad (7.1\text{a})$$

where

$Tx\ SBN$ = transmitter sideband noise that falls in the receiver's bandwidth (dBm/8 kHz);

$Tx\ SBN_O$ = normalized transmitter sideband noise at the receive frequency (dBc/Hz);

P_{TX} = transmitter power at duplexer antenna port (dBm);

B = filter pass bandwidth of the receiver (Hz).

So, in this example:

$$\begin{aligned}&Tx\ SBN\ (\text{dBm/8 kHz})\\&= Tx\ SBN_O\ (\text{dBc/Hz}) + P_{\text{TX}} + [10 \times \text{LOG}_{10}\ (B)]\\&= -150 + 50 + [10 \times \text{LOG}_{10}\ (8 \times 10^3)]\\&= -61\ \text{dBm/8 kHz}\end{aligned} \quad (7.1\text{b})$$

Now, at the receiver end, we start by calculating the receiver thermal noise floor.

$$N_O\ (\text{dBm}) = -174 + [10 \times \text{LOG}_{10}\ (B)] + Rx\ NF \quad (7.2\text{a})$$

where

N_O = thermal noise power at the duplexer antenna port (dBm);

B = filter pass bandwidth of the receiver (Hz);

$Rx\ NF$ = base station receiver noise figure (dB).

Remember, because most LMR systems operate in 12.5-kHz channels, practical LMR receivers have a filter bandwidth of around 8 kHz. Also, the noise figure of a typical LMR base station receiver is around 8 dB, so, using these numbers, we get a thermal noise floor of:

$$\begin{aligned} N_O \text{ (dBm)} &= -174 + [10 \times \text{LOG}_{10} (8 \times 10^3)] + 8 \\ &= -127 \text{ dBm/8 kHz} \end{aligned} \tag{7.2b}$$

At this point, before calculating the interference threshold, we would normally add the site noise. However, in this case, let us assume there is no site noise as this results in the highest required transmit to antenna filter rejection (because the receiver noise floor is at its lowest possible value). Hence, we can say that:

$$N_X = 10 \times \text{LOG}_{10} (N_T + I_T) \tag{7.3}$$

where

N_X = total noise and interference floor referred to the duplexer antenna port (dBm);

N_T = total noise power at the duplexer antenna port (mW); note that in this example, N_T is the same value as N_O because we are assuming there is no additional site noise;

I_T = interference threshold, the maximum permissible level of transmitter sideband noise at the duplexer antenna port (mW).

So, to find the maximum permissible level of transmitter sideband noise at the duplexer antenna port (in other words, the interference threshold, I_T), the equation becomes:

$$I_T = 10 \times \text{LOG}_{10} (N_X - N_T) \tag{7.4a}$$

where

I_T = interference threshold, the maximum permissible level of transmitter sideband noise at the duplexer antenna port (dBm);

N_X = total noise and interference floor referred to the duplexer antenna port (mW);

N_T = total noise power at the duplexer antenna port (mW).

Again, remember the units of the noise levels must be converted to milliwatts first. So, going back to the calculation, if the level of transmitter sideband noise (I_T) that we can add to the −127-dBm thermal noise floor (N_O) causes the sum (N_X) to increase by 1 dB, we can say that:

$$\begin{aligned} N_X &= N_T + 1 \text{ dB} \\ &= -127 + 1 \\ &= -126 \text{ dBm} \end{aligned} \tag{7.4b}$$

So, converting to milliwatts:

$$\begin{aligned} N_X\,(\text{mW}) &= 10^{\,(N_X\,(\text{dBm})/10)} \\ &= 10^{(-126/10)} \\ &= 2.51 \times 10^{-13} \text{ mW} \end{aligned} \tag{7.4c}$$

and

$$\begin{aligned} N_T\,(\text{mW}) &= 10^{(N_T\,(\text{dBm})/10)} \\ &= 10^{(-127/10)} \\ &= 2 \times 10^{-13} \text{ mW} \end{aligned} \tag{7.4d}$$

Then, by substituting in these values, we find the maximum permissible level of transmitter sideband noise (I_T):

$$\begin{aligned} I_T &= 10 \times \text{LOG}_{10}\,(N_X - N_T) \\ &= 10 \times \text{LOG}_{10}\,[(2.51 \times 10^{-13}) - (2 \times 10^{-13})] \\ &= -132.87 \text{ dBm} \end{aligned} \tag{7.4e}$$

As a useful rule of thumb, if you take a noise source in dBm and add another noise source 6 dB lower, the total noise increases by 1 dB. Hence, the interference threshold that causes the 1-dB increase (−132.9 dBm in this case) is about 6 dB lower than the receiver noise floor (−127 dBm).

At this point, we need to jump ahead of ourselves briefly and consider that, as well as transmitter sideband noise, the base station receiver connected to the duplexer is also exposed to the remnant of the transmitted signal that passes through the duplexer antenna to receive port filter and could cause receiver blocking. In effect, then, there are two interference mechanisms at play, and the interference threshold must be modified accordingly so that the total interference only results in 1-dB loss of receiver sensitivity. Revising from Chapter 6, if multiple interference mechanisms are present:

$$\text{Interference per mechanism (dBm)}, \; I'_T = I_T - [10 \times \text{LOG}_{10}\,(n)] \tag{7.5a}$$

where

I_T = interference threshold, the total interference level at the base of the Rx antenna (dBm);

n = number of interference mechanisms simultaneously occurring.

So, in this case:

$$\begin{aligned} \text{Interference per mechanism (dBm), } I'_T &= I_T - [10 \times \text{LOG}_{10}\ (n)] \\ &= -132.87 - [10 \times \text{LOG}_{10}\ (2)] \\ &= -135.87 \text{ dBm} \end{aligned} \quad (7.5b)$$

Intuitively, if two interference mechanisms are present, then the interference per mechanism can only be half the permissible total.

We have now calculated the actual transmitter sideband noise level at the duplexer antenna port on the receive frequency (*Tx SBN*), and the maximum permissible level (I'_T) that we can tolerate if we are only going to allow 1 dB of receiver desensitization when the transmitter is operating. So, the required sideband noise rejection in the duplexer transmit to antenna filter is simply the difference between the two values:

$$\begin{aligned} &\text{Transmitter sideband noise rejection at receive frequency (dB)} \\ &= Tx\ SBN - I'_T \\ &= -61 - (-135.87) \\ &= 75 \text{ dB} \end{aligned} \quad (7.6)$$

It is more valuable perhaps to represent this on a diagram such as Figure 7.5 as it tends to bring the calculation to life.

7.2.2 Antenna to Receiver Filter

Moving on now, consider the other filter, between the antenna and receive ports of the duplexer. This filter passes the desired signal to the receiver while rejecting the transmit signal and any unwanted received signals. Obviously, in the case of duplexed RF systems, the full transmit signal is present on the duplexer antenna port shared by the receiver, so, first and foremost, the antenna port to receive port filter must attenuate the transmit signal enough to prevent the base station receiver being damaged. If we carry on with our previous example (where the level of the transmitter signal at the duplexer antenna port was +50 dBm) and assume that a typical base station receiver front end can survive prolonged exposure to unwanted signals of +10 dBm, then the duplexer antenna to receive port filter must offer at least 40 dB of rejection at the transmit frequency. That isn't likely to be enough though, as

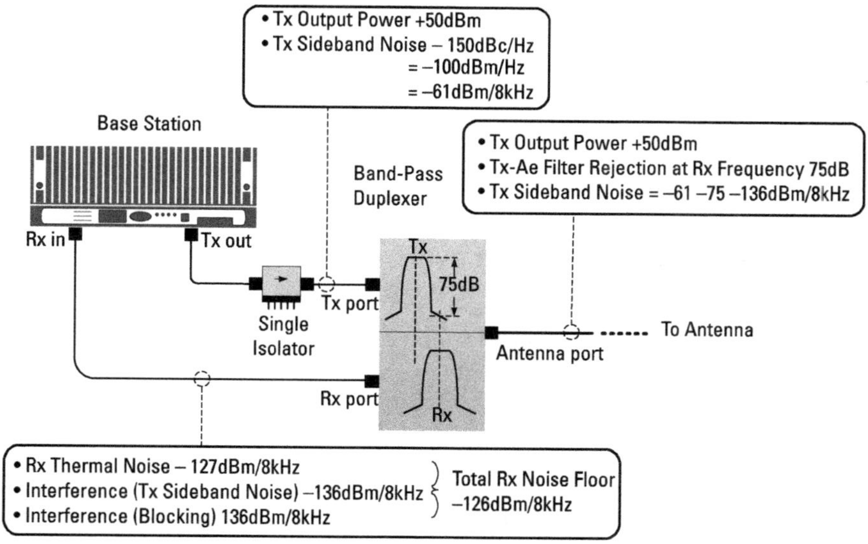

Figure 7.5 Duplexer transmitter sideband noise suppression.

the level of the transmit signal reaching the receiver will still be strong enough to cause receiver blocking, that is, affect the base station receiver circuitry in such a way as it causes a reduction in sensitivity.

Generally, the blocking performance of a good quality base station receiver is limited to around 105 dB by the reciprocal mixing mechanism whereby the remains of the transmit signal passes through the receiver's front end to the mixer RF port and transposes local oscillator noise onto the receiver intermediate frequency (IF) circuitry. However, the 105-dB blocking performance figure is a lab measurement made with a wanted signal and an interfering signal that are static. In practice, the wanted signal (weak uplink signal from a user in the field) will experience Rayleigh fading, whereas the interfering signal will not, as it is coming from a colocated transmit antenna within line of sight. To account for this, we must reduce the blocking performance by 7.2 dB, an amount equal to the difference between the static carrier-to-noise ratio ($C_S\ /N$), which was 8 dB, and the dynamic Rayleigh faded carrier-to-noise and interference ratio ($C_F\ /(N+I)$), which was 15.2 dB. The blocking performance hence becomes 97.8 dB. The base station receiver is also subject to sideband noise from the transmitter sharing the duplexer, so, in effect, there are two on-channel noise sources present. We can account for this by reducing the blocking performance by another 3 dB to 94.8 dB. What that means is that the maximum transmit signal level that we can tolerate at the duplexer receive port is 94.8 dB above the minimum wanted receive signal.

The next step is to calculate the best possible dynamic receiver sensitivity. The calculation starts with the receiver thermal noise floor (N_O).

$$\text{Receiver thermal noise floor, } N_O \text{ (dBm)} = -174 + [10 \times \text{LOG}_{10}(B)] + Rx\ NF \tag{7.7a}$$

where

N_O = thermal noise power at the duplexer antenna port (dBm);

B = filter pass bandwidth of the receiver (Hz);

$Rx\ NF$ = base station receiver noise figure (dB).

Remembering that practical LMR receivers have a filter bandwidth of around 8 kHz and the noise figure of a typical LMR base station receiver is around 8 dB, we get a thermal noise floor of:

$$\begin{aligned} N_O \text{ (dBm)} &= -174 + [10 \times \text{LOG}_{10}(8 \times 10^3)] + 8 \\ &= -127 \text{ dBm/8 kHz} \end{aligned} \tag{7.7b}$$

To this, we add 1 dB, as that is the allowable desensitization, so the total noise floor (N_x) becomes:

$$N_X = N_T + 1 \text{ dB} \tag{7.8a}$$

where

N_X = total noise and interference floor referred to the duplexer antenna port (dBm);

$N_T = N_O$ = thermal noise power at the duplexer antenna port (mW).

So

$$\begin{aligned} N_X &= N_T + 1 \text{ dB} \\ &= -127 + 1 \\ &= -126 \text{ dBm} \end{aligned} \tag{7.8b}$$

To find the receiver sensitivity, we add the relevant $C_F\ /(N + I)$ threshold to the total noise floor (N_O); the $C_F\ /(N + I)$ threshold being how far the received signal must be above the noise and interference in a real-world Rayleigh faded environment in order to provide the minimum acceptable level of link performance. So for our P25 Phase 1 system using C4FM modulation where a minimum DAQ of 3.0 is required, the receiver sensitivity becomes:

$$\begin{aligned}\text{Dynamic receiver sensitivity (dBm)} &= N_X + [C_F\ /(N+I)] \\ &= -126 + 15.2 \\ &= -110.8\text{ dBm}\end{aligned} \tag{7.9}$$

where

Dynamic receiver sensitivity = the received signal level required for minimum performance (dBm);

N_X = total noise and interference floor referred to the duplexer antenna port (dBm);

$C_F\ /(N+I)$ = the Rayleigh faded carrier-to-noise and interference ratio (dB), which, in this case, is 15.2 dB (see Chapter 6, Table 6.1).

This is the best the receiver sensitivity can be for this site receiving system because the calculation only includes thermal noise and not any additional site or environmental noise.

The maximum transmit signal level that we can therefore tolerate then at the duplexer receive port is the sum of the above, the minimum wanted receive signal, and the blocking performance:

$$\begin{aligned}&\text{Maximum transmit signal at receive port (dBm)} \\ &= \text{Receiver sensitivity} + \text{Blocking} \\ &= -110.8 + 94.8 \\ &= -16\text{ dBm}\end{aligned} \tag{7.10}$$

We now know both the transmitter signal level at the duplexer antenna port (+50 dBm) and the maximum permissible level that we can tolerate at the duplexer receive port (−16 dBm if we are only going to allow 1 dB of receiver desensitization when the transmitter is operating). So, the rejection required in the duplexer antenna to receive filter at the transmit frequency is simply the difference between the two:

$$\begin{aligned}&\text{Duplexer antenna to receive rejection at transmit frequency (dB)} \\ &= \text{Transmit signal level at duplexer antenna port (dBm)} \\ &\quad - \text{Maximum transmit signal level at receive port (dBm)} \\ &= 50 - (-16) \\ &= 66\text{ dB}\end{aligned} \tag{7.11}$$

Again, to bring some life to the calculation, here it is in Figure 7.6 represented on a diagram.

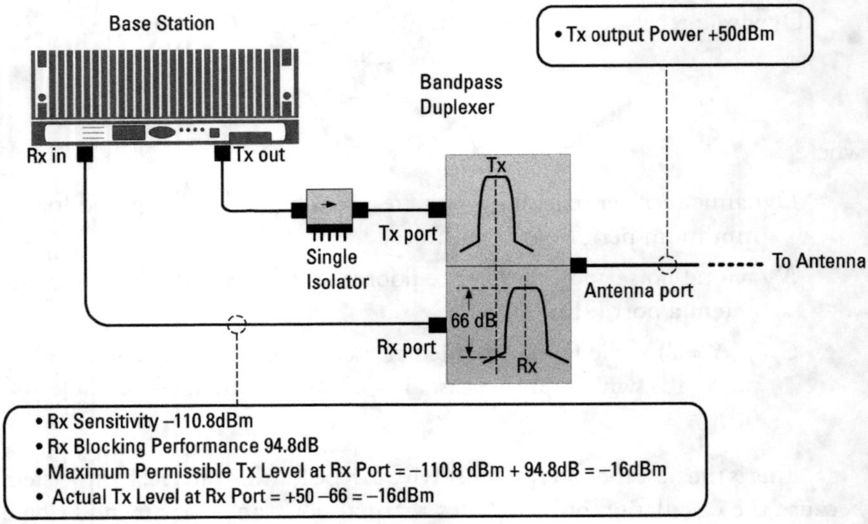

Figure 7.6 Duplexer receiver blocking rejection.

Note that the duplexer insertion loss and cable losses have been deliberately ignored in all the calculations; in single-channel site RF systems, these losses are small in the big scheme of things and just clutter up the calculations. In effect, ignoring them simply means that you have a few extra decibels in your back pocket.

Also note that these calculations show how transmitter sideband noise suppression and receiver blocking rejection are derived for a single-channel site RF system with none of the other interference mechanisms present. They will serve as a starting point to be expanded upon in Chapters 8 and 9, where we will consider how to do similar calculations for multichannel systems.

7.2.3 Final Duplexer Frequency Response

To wrap up these calculations pictorially then, Figure 7.7 shows what a frequency response plot of our example duplexer would look like. Note that an extra 5 dB has been added to both the calculated rejection figures to account for temperature drift and aging.

Duplexer vendors will have a range of standard models, but it is up to us as the engineers to specify the performance that we need, usually in the form of Table 7.1. The Tx port to antenna port and the antenna port to Rx port rejection figures in the table are those from our example, while typical figures

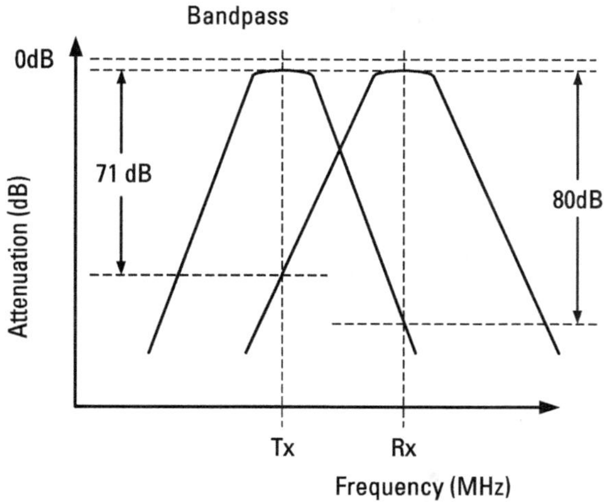

Figure 7.7 Duplexer frequency response plot.

Table 7.1
Example Duplexer Specification

Parameter	Design Limit
Filter characteristic	Bandpass
Tuneable frequency range, Tx port	165 +/–2 MHz
Tuneable frequency range, Rx port	170 +/–2 MHz
Tx–Rx frequency separation (typical)	≥4.6 MHz
Tx port to antenna port insertion loss	≤1.0 dB
Tx port to antenna port passband	≥1 MHz
Tx port return loss in Tx passband	Better than 22 dB
Single-carrier CW RF input power	≥100W (+50 dBm)
Antenna to Rx port insertion loss	≤1.5 dB
Antenna port to Rx port passband	≥1 MHz
Antenna port return loss in Rx passband	Better than 18 dB
Noise rejection at Rx frequency, Tx port–antenna port	≥80 dB
Rejection at Tx frequency, antenna port–Rx port	≥71 dB
RF connectors	Type N Female

are given for the remaining parameters. In many cases, the chosen duplexer vendor will have a standard model that meets the requirements.

7.3 Duplexers and Passive IM

Before finishing the subject of duplexers, there is one more point to cover. For single-channel repeater site RF systems with one transmitter and one receiver, the use of one antenna and a duplexer is recommended. However, for multichannel systems, with several transmit and receive channels, it is not quite that simple. To see why, let us start with the diagram of a two-channel duplexed system shown in Figure 7.8, because it will simplify the following explanation.

The problem with multichannel duplexed systems becomes apparent when we consider transmitter IM. Consider the following VHF example where we have two combined transmit channels at 150.0 and 150.5 MHz and two related receive channels at 146.0 and 146.5 MHz. Once the transmitters are combined, any nonlinearities in the duplexer or discontinuities in the RF cabling to the antenna will cause the transmit signals to mix and generate

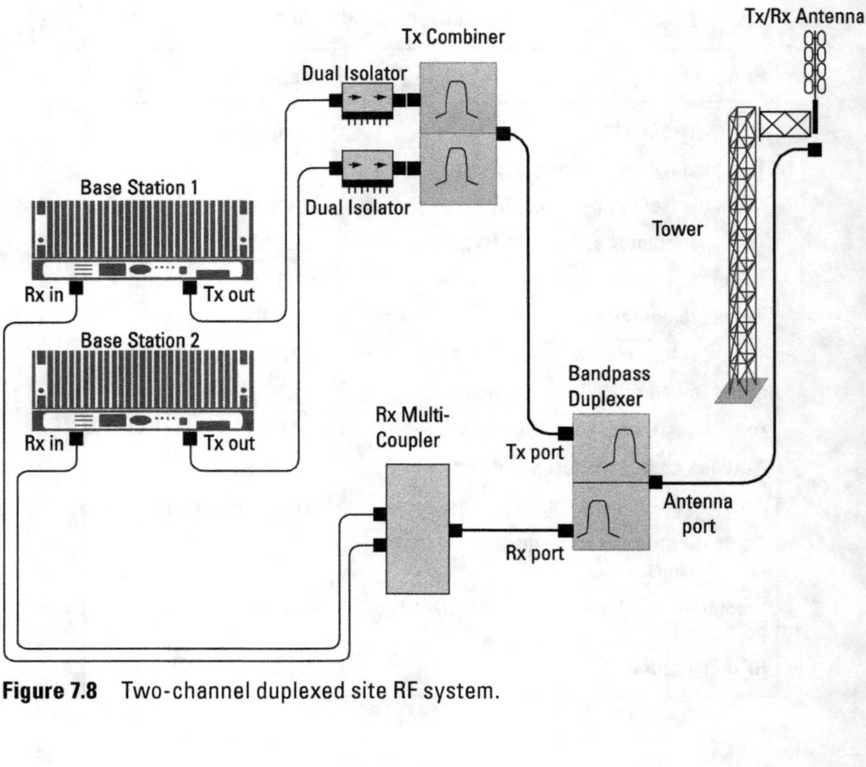

Figure 7.8 Two-channel duplexed site RF system.

IM products on the antenna side of the duplexer as shown in Figure 7.9. This phenomenon is known as passive IM.

Corroded RF connectors or damaged RF cables are classic causes of this effect and are real risks because they are outside; and the worse the corrosion or damage gets over time, the worse the passive IM becomes.

The thing to note from the example shown in Figure 7.9 is the fifteenth-order and seventeenth-order IM products are generated at exactly the same frequencies as the receive channels (146.5 MHz and 146.0 MHz, respectively). They are in the passband of the duplexer antenna to the receive filter and pass straight through to the multicoupler and the connected receivers. If the amplitude of these passive IM products is greater than the noise floor of the receiver, then the receiver sensitivity is degraded leading to a loss of coverage.

Wherever possible then, if you have to use a duplexer in a multichannel site RF system (and there are times when you have to, perhaps due to loading on the tower), try only to combine transmit channels that do not generate IM products on a receive channel connected to the same antenna. The best way to go about this is not to combine transmit channels where the transmit to receive frequency separation is an integer multiple of the frequency separation between the transmit channels.

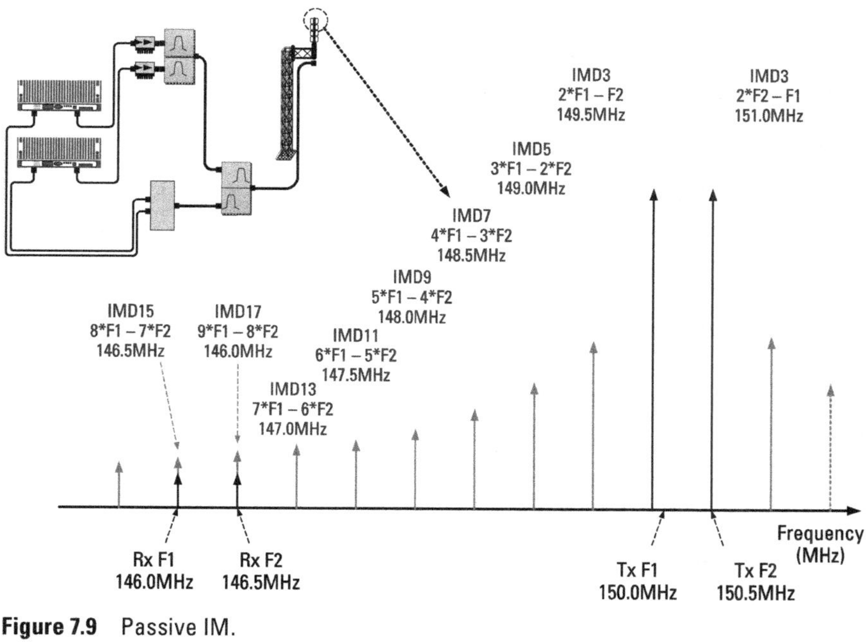

Figure 7.9 Passive IM.

However, now think about what would happen if we were to use separate transmit and receive antennas instead, as shown in Figure 7.10.

Passive IM still occurs on the antenna side of the transmitter combiner, as the RF connectors and cables are subject to the same corrosion and damage as they would be in a duplexed antenna system. However, there is no longer a direct path for those on channel fifteenth-order and seventeenth-order passive IM products to reach the receiver. Instead, the isolation between the transmit and receive antennas provides additional protection. If we were to achieve the recommended 60-dB isolation between the transmit and receive antennas, the level of the passive IM products reaching the receive antenna would drop by 60 dB too, and most of them would disappear into the noise floor as shown in Figure 7.11.

Good rules to adhere to then, as far as you can, are:

- For single-channel site RF systems, use a duplexer and one antenna.
- For multichannel site RF systems, use separate transmit and receive antennas with a transmitter combiner and receiver multicoupler.

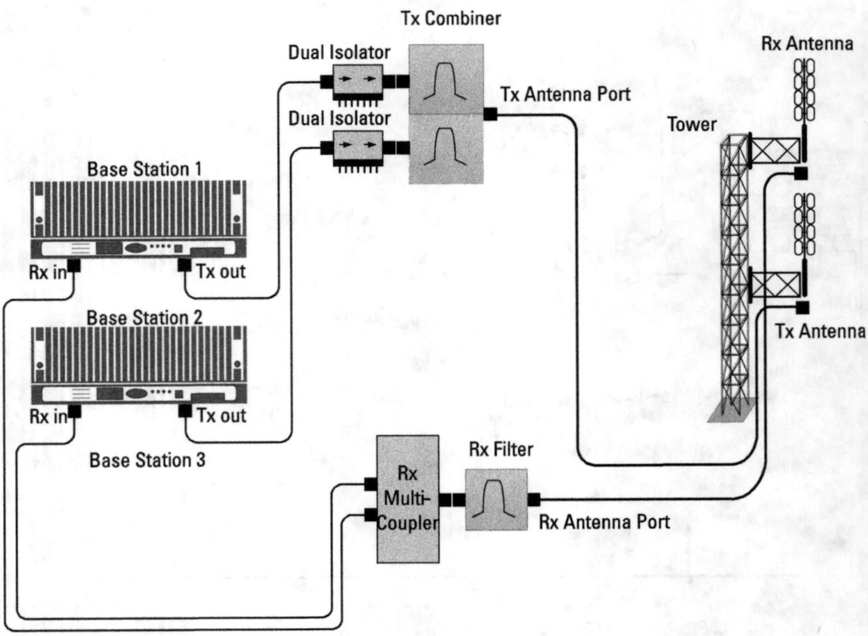

Figure 7.10 Two-channel site RF system.

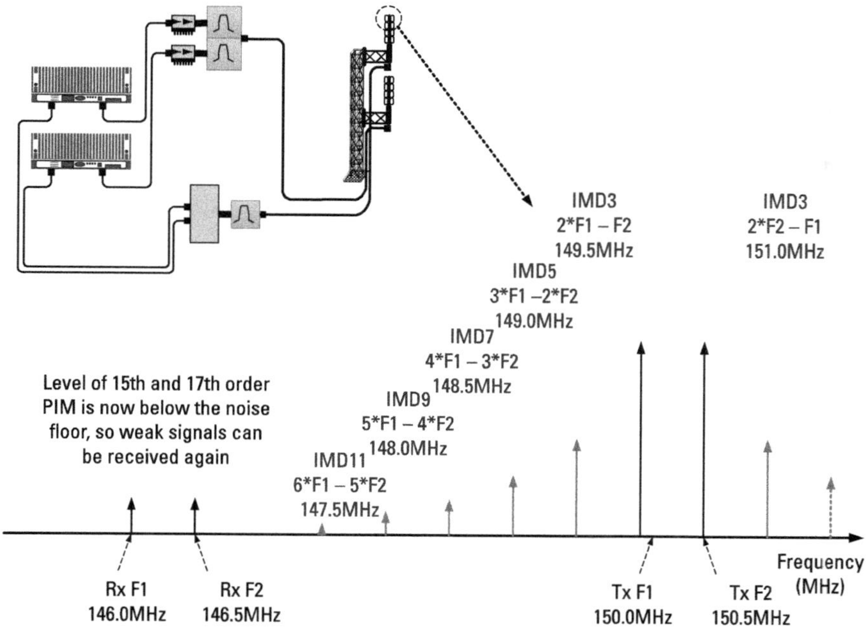

Figure 7.11 Passive IM with separate transmit and receive antennas.

Finally, note that while the frequencies at which passive IM may occur in any given multichannel duplexed site RF system are easy to predict, the levels are not. This is because the levels of passive IM products do not follow the rules presented in Chapter 3. In some cases, decreasing the levels of the contributing signals can lead to an increase in the level of the passive IM product.

Selected Bibliography

Comprod 66-38-74. Cavity Filter Duplexer, https://comprodcom.com/shop/filters-rf-components/duplexers/pseudo-band-pass-duplexer-66-38-74/.

dBSpectra, MWFDP9AM Milled Window Filter Duplexer, https://www.dbspectra.com/products/896-940-mhz-milled-window-filter-duplexer-mwfdp9am/.

Telecommunications Industry Association, Report TSB-88.1-F, Wireless Communications Systems—Performance in Noise and Interference Limited Situations—Part 1: Recommended Methods for Technology Independent Narrowband Performance Modelling.

8

Transmitter Combining Systems

A transmitter combiner allows a number of base station transmitters to be connected to a common transmit antenna, hence minimizing real estate and loading requirements on the tower. The design of the transmitter combiners for each site in the network (and they can all be different) can only be started after the site survey, tower mapping, and spectrum and noise monitoring have been completed, subsequent frequency planning has been carried out, and the antenna placements on the tower have been finalized.

Most transmitter combiners consist of a number of highly selective bandpass cavity filters, at least one for each transmitter. Each filter is tuned to the transmit frequency of that particular base station and the filter outputs are then connected together (combined) in one of two ways with coaxial cables. Before the cavity filters, isolators are connected to the transmitter output of each base station to minimize generation of transmitter IM products within the combiner and provide protection in the event of a fault elsewhere in the combining system. There is a point where cavity filter-based transmitter combiners cannot be used anymore if the frequencies of the transmitters become too close together; in this situation, hybrid combiners based on 3-dB directional couplers are used instead.

Figure 8.1 is a reminder of where the transmitter combiner resides in a multichannel repeater site RF system.

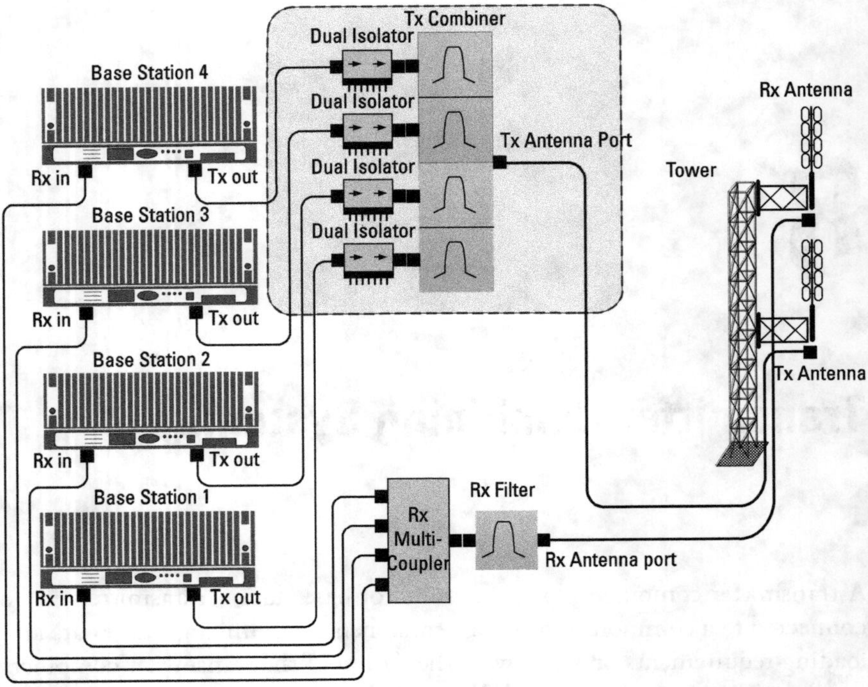

Figure 8.1 Use of a transmitter combiner in a multichannel site RF system.

In this chapter, you will learn how the various types of transmitter combiner work so that you can choose the correct type for each site in your system and then learn how to specify the design requirements. Before that, though, some background on isolators and cavity filters is necessary. You do not have to physically design and make these yourself; like the duplexers in Chapter 7, you buy them from your chosen vendor, but you do need to know enough to understand how the various types of transmitter combiners work.

8.1 Isolators

To start, let us look at the difference between a circulator and an isolator. A circulator is a three-port passive ferrite device that only allows a signal to pass between neighboring ports in one direction. For example, in Figure 8.2, the signal can pass at low loss between ports 1 and 2, ports 2 and 3, or ports 3 and 1. However, it cannot pass from port 2 to port 1, port 3 to port 2, or port 1 to port 3.

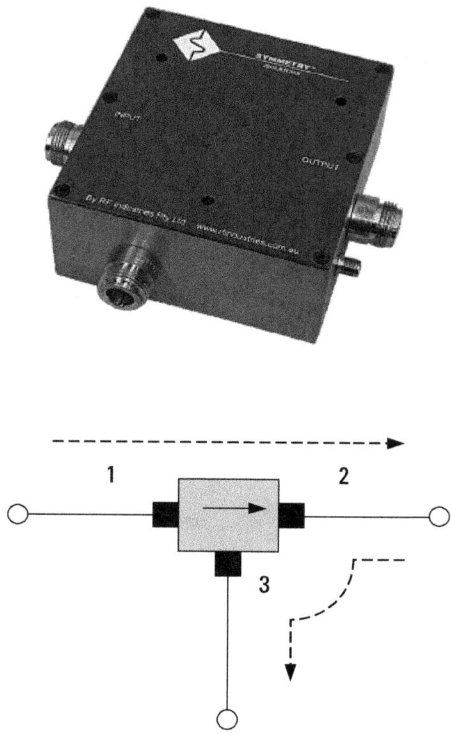

Figure 8.2 Circulator.

When port 3 is terminated in a high-power, 50-ohm load as in Figure 8.3, this property becomes particularly useful. It means that signals can pass from port 1 (input) to port 2 (output) at a low loss, but a signal coming the other way from port 2 cannot reach the port 1; instead, it ends up in the 50-ohm load at what was port 3. In this application, a circulator becomes known as an isolator, basically because it isolates the output (port 2) from the input (port 1).

Isolators are therefore very useful in transmitter combining systems for:

- Minimizing the generation of transmitter IM products by preventing the signal from one transmit channel reaching the output of another transmitter on the combiner;
- Ruggedizing the transmitter system by protecting it from reverse power reflected by a fault elsewhere in the following transmitter combining system.

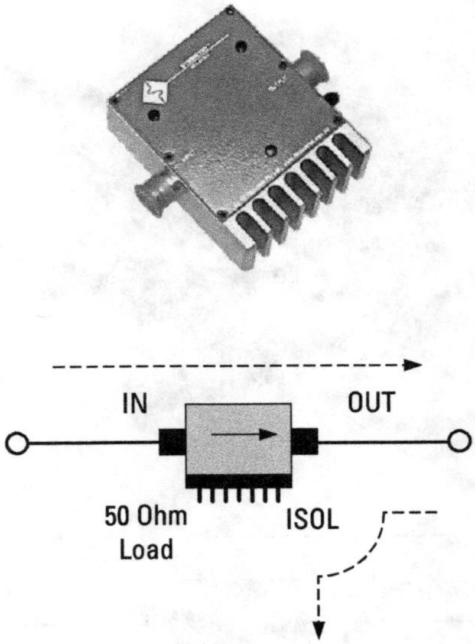

Figure 8.3 Isolator.

Typical isolator specifications are given in Table 8.1, according to the band of use.

An important thing to note in Table 8.1 is that the tunable bandwidth is the bandwidth over which the isolator may be tuned. However, because isolators usually have only a single-stage tuning circuit in each port, the tuned bandwidth is the frequency interval over which the isolator can be tuned at

Table 8.1
Example Isolator Specification

Parameter	Typical Isolator Performance			
Frequency band	66–88 MHz	VHF	UHF	800 MHz
CW RF input power	≥100W	≥100W	≥100W	≥100W
Tunable bandwidth (MHz)	20	30	40	50
Tuned bandwidth (MHz)	F_c+/−1	F_c+/−2	F_c+/−5	F_c+/−10
Power rating (W)	50	150	150	150
Insertion loss (dB)	0.5	0.3	0.3	0.3
Return loss (dB)	>22	>22	>22	>22
Isolation (dB)	>30	>35	>35	>35

any one time. This basically means that isolators need to be tuned specifically to the output frequency of the transmitter to which they are connected.

One last thing to mention is that the isolation provided by a single isolator stage is often not enough to suppress the generation of IM products in transmitter combiners to levels low enough to not interfere with colocated site receivers. For this reason, dual isolators (two single isolators connected in series inside one case) like that shown in Figure 8.4 are readily available. In terms of specifications, dual isolators have twice the isolation but twice the insertion loss of single isolators.

Note that the load must be capable of dissipating all the RF power from the connected transmitter in the event of a fault, as 100% reflection could occur in extreme cases.

8.2 Cavity Filters

Cavity filters like those in Figure 8.5 are mechanically tunable resonant circuits that pass a narrow band of frequencies and reject all others (bandpass)

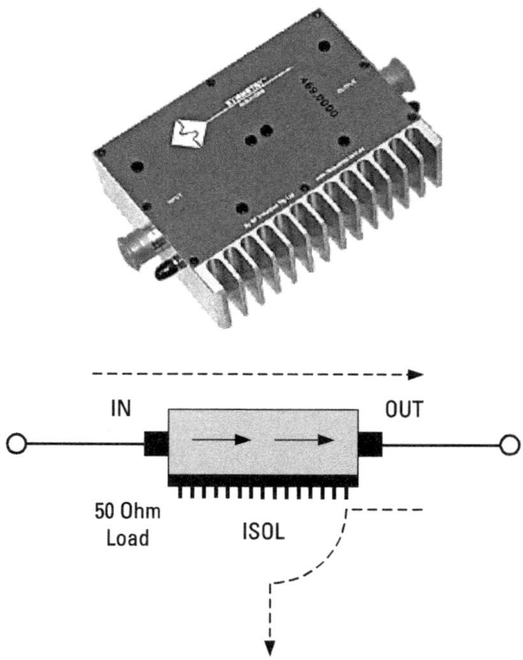

Figure 8.4 Dual isolator.

Figure 8.5 Cavity filters.

or reject a specific frequency and pass all others (band-reject). Because of the mechanical construction, they are very selective; at LMR RF, bandpass cavity filters can reject unwanted noise or signals only a megahertz or so away.

If we took the top plate off a bandpass cavity filter and looked inside the can to find out how it worked (as engineers do), we would see it consisted of an adjustable $\lambda/4$ or $3\ \lambda/4$ resonator and separate rotatable input and output coupling loop assemblies as shown in Figure 8.6.

Bandpass cavity filters are the most commonly used type, being tuned to pass a very narrow band of frequencies and attenuate anything outside that band. They are used in transmitter combiners to pass the desired transmit frequencies while cleaning up the associated sideband noise skirts to prevent desensitization of nearby receivers at the site.

The desired center frequency is set by adjusting the $\lambda/4$ resonator to the right electrical length with the tuning plunger, after which the coupling loops are rotated to set the required bandwidth and rejection characteristic. Figure 8.7 shows that there is an inherent trade-off between the bandwidth and the insertion loss of a bandpass cavity filter; you do not get anything for nothing, so the smaller the pass bandwidth, the greater the insertion loss becomes. For example, when the cavity filter coupling loops are adjusted so that the insertion loss is 1 dB, the 30-dB bandwidth will typically be around

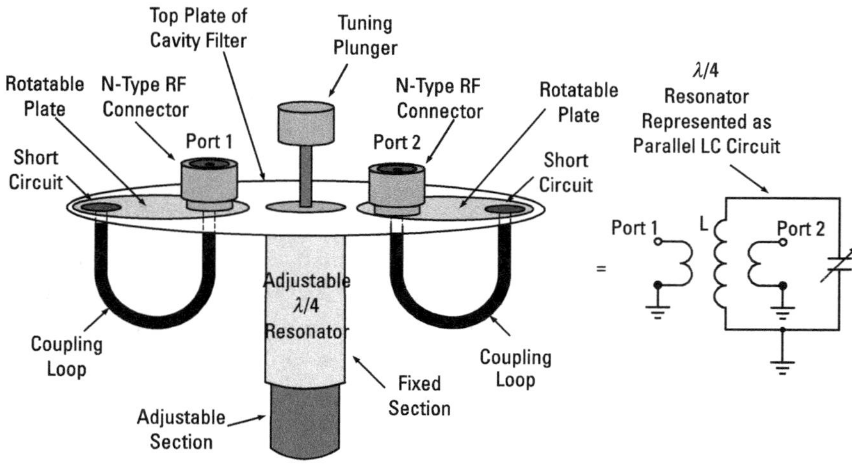

Figure 8.6 Cavity filter construction and equivalent circuit.

F_c+/–5 MHz. If the loops are rotated so the coupling becomes greater, the insertion loss might drop to 0.5 dB, but the 30-dB bandwidth will widen to around F_c+/–10 MHz. Conversely, with the loops adjusted for lower coupling, the insertion loss increases to around 2 dB but the 30-dB bandwidth can be narrowed to as little as F_c+/–3 MHz.

While bandpass is the most commonly used type, cavity filters are also available in a band-reject (notch) configuration whereby the filter is tuned to reject a particular frequency and pass everything else. Band-reject cavity filters should never be used on their own in transmitter combiners. Instead, they

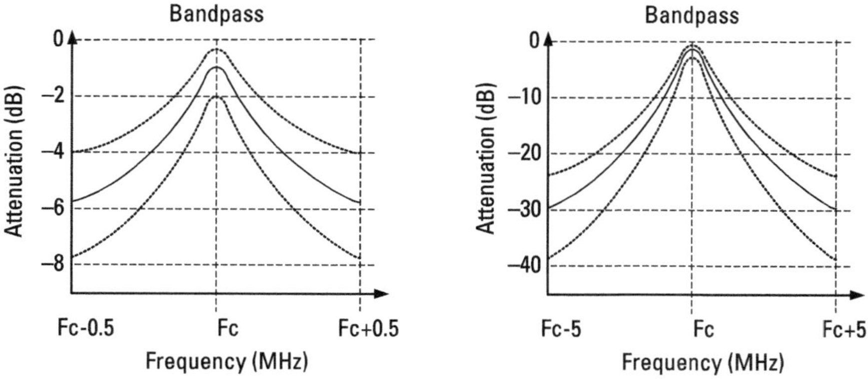

Figure 8.7 Bandpass cavity filter frequency response.

are used together with bandpass cavities, often in the arm of a transmitter combiner that happens to be very close in frequency to that of a receiver at the site. The notch is tuned to the nearby receiver's frequency to specifically clean up the transmitter sideband noise emitted at that particular frequency.

Again there is an inherent trade-off, this time between the notch depth and the bandwidth; the greater the notch depth, the wider the bandwidth. As a typical example of this, if the frequency we need to reject is within 250 kHz of the wanted (pass) frequency, then the notch depth will only be around 10 dB. If the frequency we need to reject is 500 kHz away from the wanted frequency, then notch depths of 20 dB are possible, and, with 1-MHz spacing, the notch deepens further to 30 dB, as illustrated in Figure 8.8. These figures are just examples; please refer to manufacturers' data for exact band-specific performance numbers.

There is a final type of cavity filter, known as bandpass/band-reject, that is simply a combination of the previous two. These are tuned to pass a narrow band of frequencies and reject a wider band, as shown in Figure 8.9. Again, bandpass/band-reject cavity filters should never be used on their own in transmitter combiners; they should be used together with bandpass cavities to clean up transmitter sideband noise at a close receive frequency. Compared to using band-reject (notch) cavity filters for this purpose, bandpass/band-reject filters add to the bandpass response.

The same trade-off between notch depth and the bandwidth exists; the closer the notch is to the passband, the less deep the notch will be. If the frequency that we need to reject is within 250 kHz of the wanted (pass) frequency, then the notch depth will only be around 10 dB. If the frequency that we need to reject is 500 kHz away from the wanted frequency, then

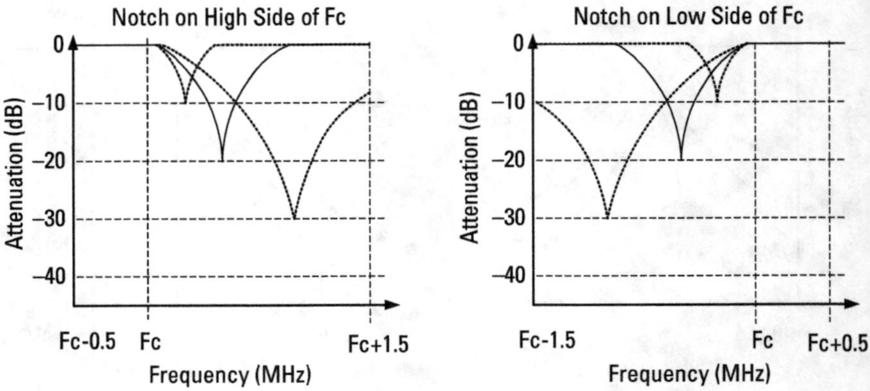

Figure 8.8 Band-reject cavity filter frequency response.

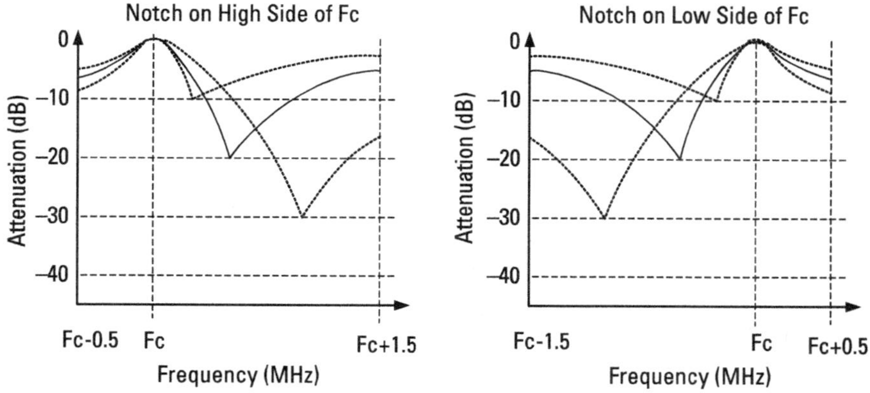

Figure 8.9 Bandpass/band-reject cavity filter frequency response.

notch depths of 20 dB are possible, and with 1-MHz spacing the notch depth increases further to 30 dB.

8.3 Types of Transmitter Combiners

Having now covered isolators and cavity filters as necessary background, let us now move on to look at the various types of transmitter combiners, how to choose the right type, and how to specify the necessary performance required at each site in the network. There are three main types of transmitter combiners: T-Pass combiner, Star combiner, and hybrid combiner. Each type has its own advantages and disadvantages, which we will examine.

8.3.1 T-Pass Transmitter Combiners

T-Pass transmitter combiners are based on cavity filters and are called such because each transmitter is connected so that it forms a T junction with the main line through to the antenna. To enable this, a modification is necessary to the output coupling loop of the cavity filter. To understand why, let us start by looking at a standard bandpass loop as shown in Figure 8.10, which is essentially a short length of transmission line connected to ground (the cavity filter body) at one end.

Figure 8.11 shows that a T-Pass loop is a variation on this theme; instead of being grounded at one end, it has another RF connector and now becomes a low-loss transmission line through the cavity filter.

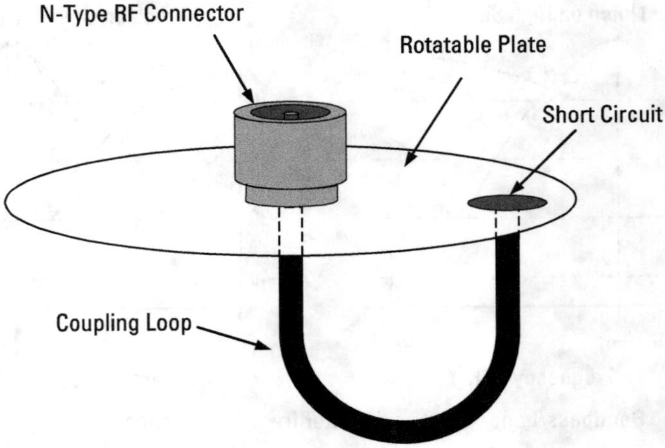

Figure 8.10 Shorted bandpass coupling loop.

Now, by attaching a short-circuit termination to the output end of the loop as shown in Figure 8.12, the T-Pass loop is converted back into a bandpass loop. This might all seem rather obvious, but bear with it and keep this thought in mind.

Next, some basic transmission line theory is necessary. A transmission line is the generic name given to a material structure that forms the path

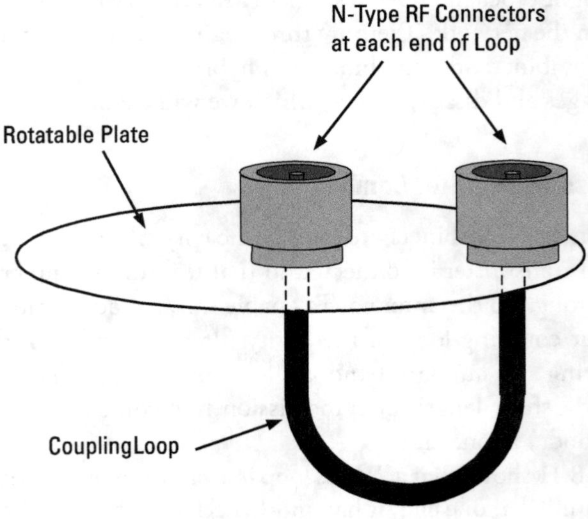

Figure 8.11 Expandable T-Pass coupling loop.

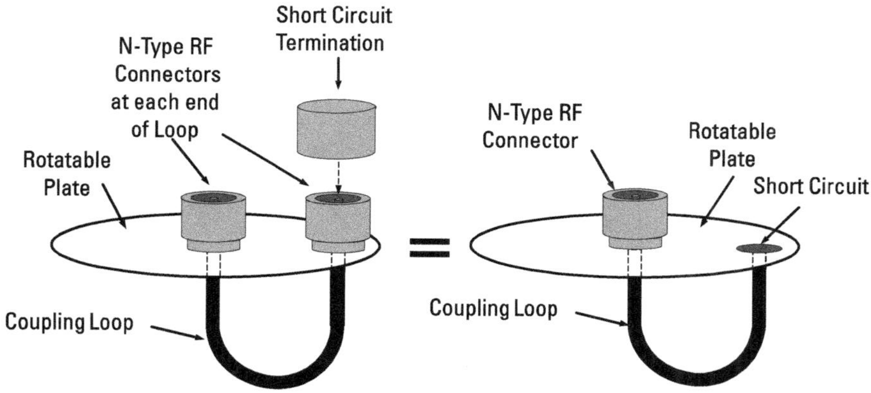

Figure 8.12 T-Pass coupling loop with short-circuit termination.

directing the transmission of energy from one place to another. Coaxial cables, waveguides, and striplines are the most common types of transmission lines used at RF. In such a transmission line, the RF current and RF voltage are 90° out of phase, so we can summarize that:

- A transmission line that is a quarter of a wavelength ($\lambda/4$) long will transform an open circuit (V = max, I = 0) to a short circuit (V = 0, I = max) or, conversely, will transform a short circuit to an open circuit, as shown in Figure 8.13.
- A transmission line that is half of a wavelength ($\lambda/2$) long will transform an open circuit to an open circuit or a short circuit to a short circuit.

If we now take the T-Pass loop and add in the transmission line theory, we can describe how T-Pass combiners work. Let us start with Figure 8.14 showing just one transmitter where the cavity filter is tuned to pass F_1 at very low loss but attenuate all other frequencies.

Now, to add a second channel, Figure 8.15 shows that all we need to do is repeat the arrangement and connect a coaxial cable (cable 1) between the outputs of T-Pass cavity filters #1 and #2 of such a length that it transforms the short-circuit termination at cavity filter #1 to appear as a virtual short circuit at the output loop of cavity filter #2. To reference this back to the transmission line theory, we need the length of cable 1 to be such that it makes the total electrical length of the connection between cavity filter #2 and the short-circuit termination at cavity filter #1 an even number of quarter-wavelengths

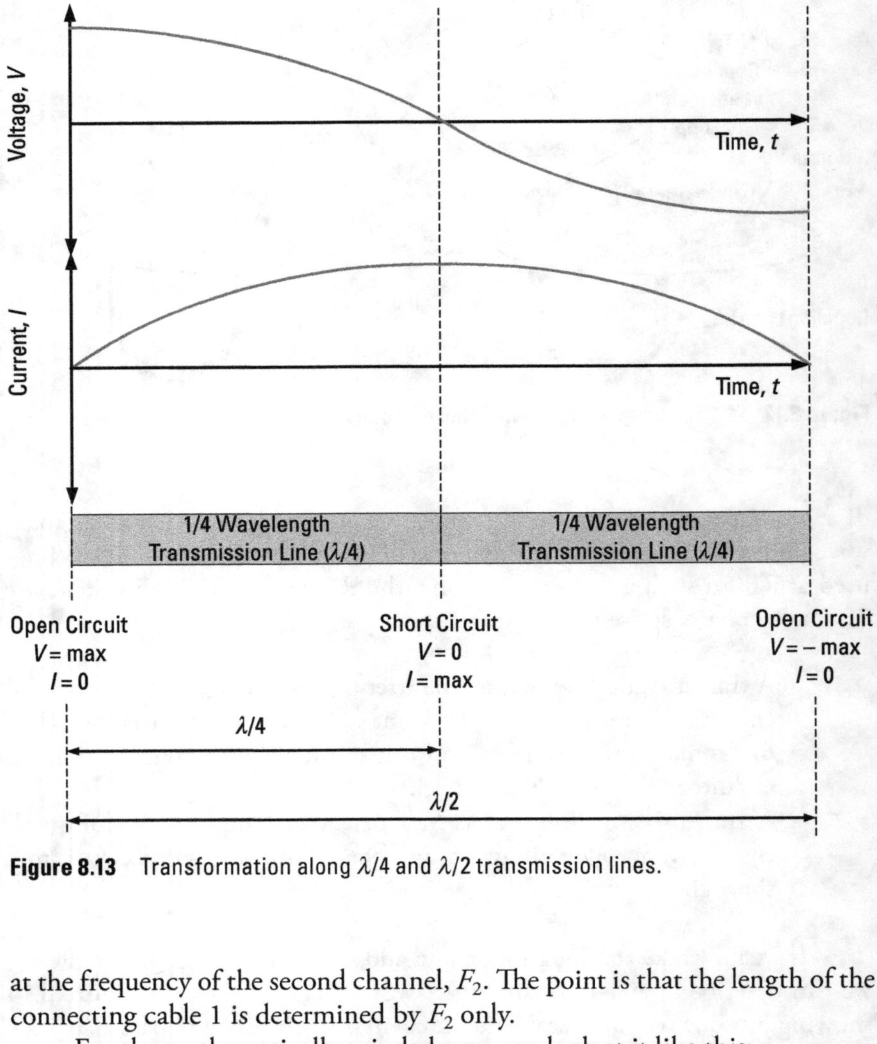

Figure 8.13 Transformation along $\lambda/4$ and $\lambda/2$ transmission lines.

at the frequency of the second channel, F_2. The point is that the length of the connecting cable 1 is determined by F_2 only.

For the mathematically minded, you can look at it like this:

$$L = \text{even number of quarter-wavelengths at } F_2 = L(C_1) + L(S_1) \quad (8.1)$$

where

L = total electrical length between connection to T-Pass cavity filter #2 and short-circuit termination at T-Pass cavity filter #1;

$L(C_1)$ = electrical length of cable 1;

$L(S_1)$ = electrical length of T-Pass loop in cavity filter #1.

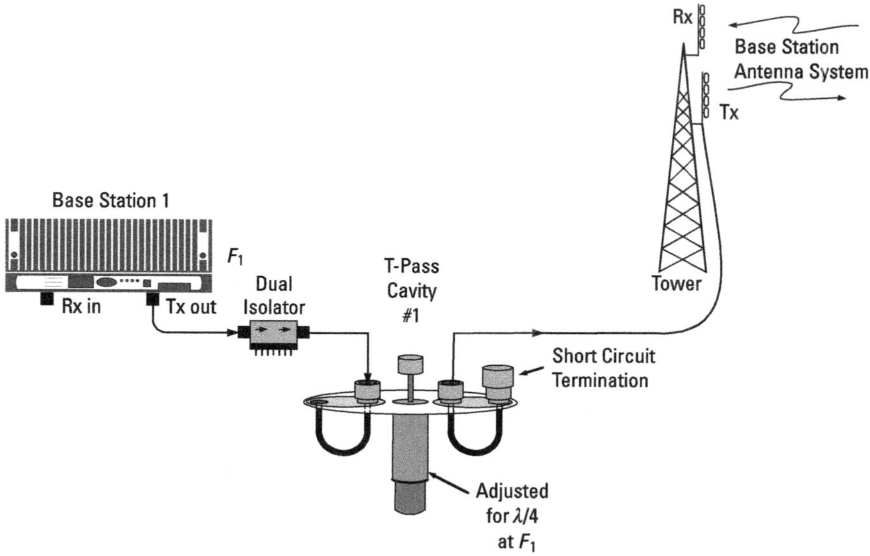

Figure 8.14 T-Pass combiner starter channel.

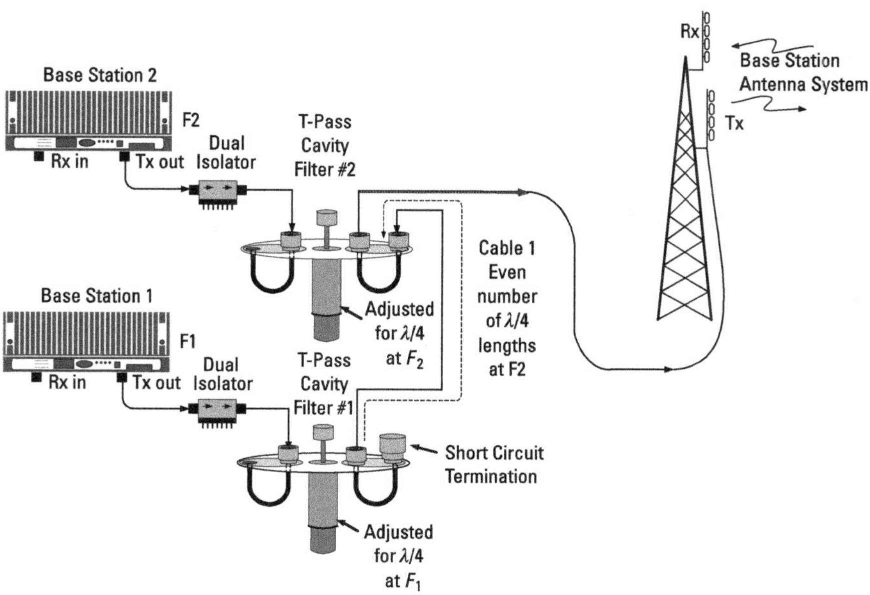

Figure 8.15 T-Pass combiner expansion channel.

The transmit frequency of base station 2 (F_2) is prevented from feeding back into the transmitter output of base station 1 by the dual isolator and by the fact that the passband of cavity filter #1 is tuned to F_1 (and thus rejects F_2, at least to some extent).

Adding a third channel is just as easy. Figure 8.16 shows that what we need to do is repeat the arrangement again and connect another coaxial cable (cable 2) between the outputs of T-Pass cavity filters #2 and #3 so that the short-circuit termination connected to cavity filter #1 appears as a virtual short circuit at the output loop of cavity filter #3. To do this, we need the length of cable 2 to be such that the total length of the connection between the output loop of cavity filter #3 and the short-circuit termination at cavity filter #1 is an even number of quarter-wavelengths at the frequency of the third channel, F_3.

Similarly, the point is that the length of the additional connecting cable 2 is determined by F_3 only. Again, for the mathematically minded:

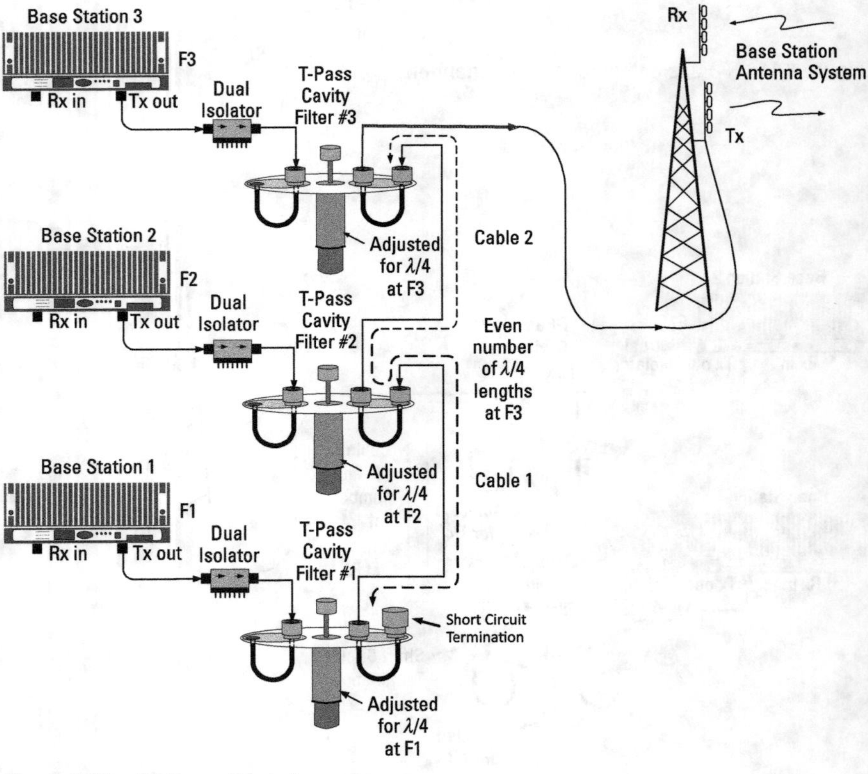

Figure 8.16 Adding a third channel.

$$L = \text{even number of quarter-wavelengths at } F_3 = L(C_2) + L(S_2) + L(C_1) + L(S_1) \quad (8.2)$$

where

L = total electrical length between connection to T-Pass cavity filter #3 and short-circuit termination at T-Pass cavity filter #1;

$L(C_2)$ = electrical length of cable 2;

$L(S_2)$ = electrical length of T-Pass loop in cavity filter #2;

$L(C_1)$ = electrical length of cable 1;

$L(S_1)$ = electrical length of T-Pass loop in cavity filter #1.

The transmit frequency of base station 3 (F_3) is prevented from feeding back into transmitter output of base stations 1 or 2 by the isolators and by the fact that the passband of cavity filter #2 is tuned to F_2 (and thus rejects F_1 and F_3). Similarly, the passband of T-Pass cavity filter #1 is tuned to F_1 (and thus rejects F_2 and F_3).

Drawn schematically, such a three-channel T-Pass transmitter combiner would look like Figure 8.17.

Insertion losses for a T-Pass combiner depend on the number of transmitters to be combined (channels) and the frequency separation between them. Typical insertion losses for a UHF T-pass combiner are as given in Table 8.2.

The losses of T-Pass combiners would be slightly less at VHF for the same transmitter frequency separation and slightly greater at 800 MHz.

In summary then, T-Pass combiners have low losses, but their main advantage is that they are easy to expand (add another channel to) after installation. All you have to do is add the extra cavity filter and one more transmission line; you do not have to touch the rest.

A couple of things to note are that the insertion loss is slightly more for the channel (F_1) at the bottom of the stack and the power is greatest at the cavity filter at the top of the stack (because the output loop is subjected to the sum of all the transmitter output powers). In designing T-Pass combiners then, we need to be careful not to exceed the power dissipation or voltage flashover limits of the top cavity filter output loop. Practically, what this does is place a limit on the number of transmitters that may be safely combined; four transmitters is a sensible number from this perspective and also from the perspective of controlling transmitter IM. If you need to combine more than four transmitters, the use of another transmit antenna with its own dedicated combiner is recommended; the transmit antennas can be at the same height

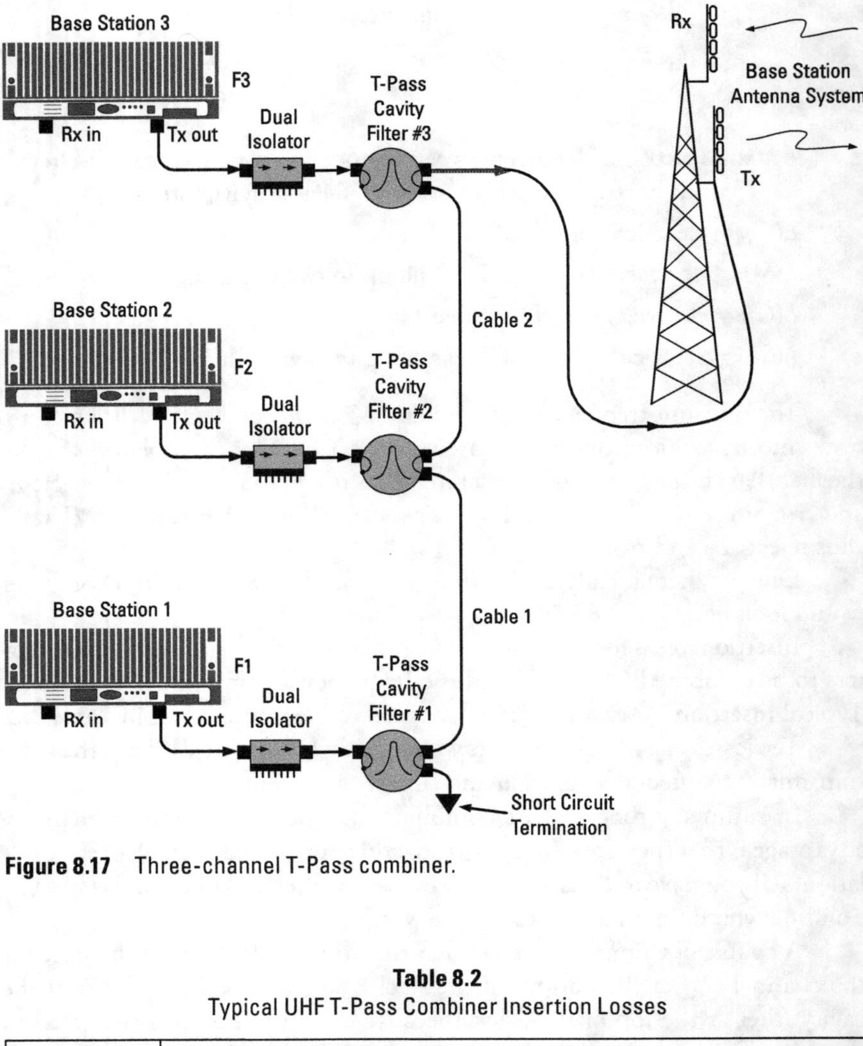

Figure 8.17 Three-channel T-Pass combiner.

Table 8.2
Typical UHF T-Pass Combiner Insertion Losses

Number of Channels	UHF T-Pass Combiner Losses at Given Tx-Tx Frequency Separation					
	100 kHz	**150 kHz**	**200 kHz**	**300 kHz**	**400 kHz**	**500 kHz**
2	4.7 dB	3.4 dB	2.7 dB	2.3 dB	2.3 dB	2.3 dB
3	4.9 dB	3.6 dB	2.9 dB	2.4 dB	2.3 dB	2.3 dB
4	5.1 dB	3.8 dB	3.1 dB	2.6 dB	2.3 dB	2.3 dB

on the tower, with the receive antenna above them as per the antenna system design rules given in Chapter 5.

8.3.2 Star Combiners

Star combiners like those shown in Figure 8.18 are the common alternative to T-Pass. They are called this because they bring all the transmitter outputs together at a common point (the Star junction) using coaxial cables that are an odd number of quarter-wavelengths to transform the short circuit at the cavity filter output loop to look like an open circuit at the junction. Unlike T-Pass combiners, no modification is necessary to the cavity filter output loops.

However, there has to be a compromise in the cable lengths, because if cable 1 from bandpass cavity filter 1 to the Star junction was a quarter-wavelength long at F_1 and cable 2 from bandpass cavity 2 to the Star junction was a quarter-wavelength long at F_2, then that cable back to bandpass cavity filter 2 would not be a quarter-wavelength at F_1. Hence, a higher level of the signal F_1 would appear at the output of bandpass cavity filter 2. So, the compromise is that all coaxial cables to the Star junction are the same length (L), which is chosen to be the average of the quarter-wavelengths of the individual transmitter frequencies. The effect of this shows up in the insertion loss of the combiner. If the difference between the highest and lowest transmitter frequencies on the combiner is small (around 1 MHz), then the insertion loss

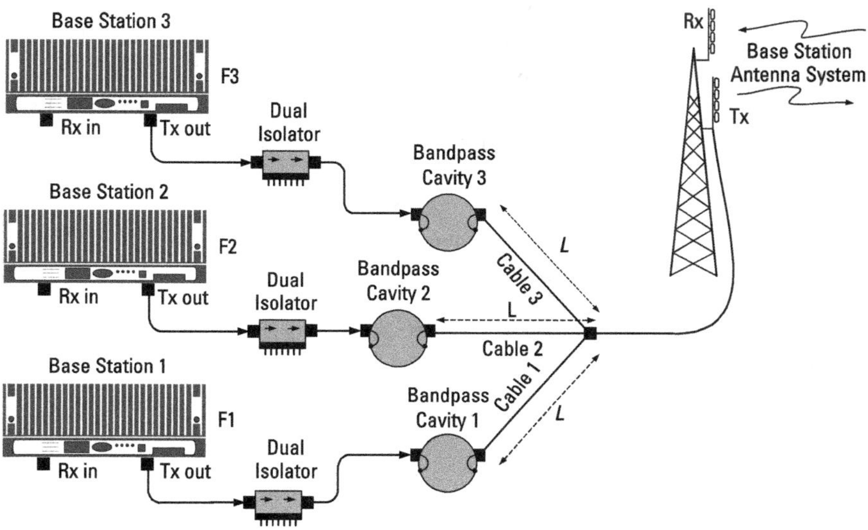

Figure 8.18 Star combiner.

per channel will be slightly lower than that of a T-Pass combiner with the same number of transmitters. However, if the difference between the highest and lowest transmitter frequencies on the combiner is wider, then the insertion loss will be slightly higher than the corresponding T-Pass counterpart.

There are a number of other key differences between T-Pass and Star combiners. First, if the need to add another channel arises after installation, then all the combining cables from the cavity filters to the Star junction need changing because the average length will now be different. It is not just a case of adding one cable and leaving the rest as they were for the T-Pass combiner. However, in reality the inconvenience of this is minimized because most vendors that favor star combiners will supply a complete new kit of connecting cables, making the task of expansion much easier.

Also, unlike T-Pass combiners, the insertion losses of Star combiners does not necessarily increase as you move down the stack, and the output loops of the cavity filters are only exposed to the RF power and voltage from the individual transmitter to which they are connected.

8.3.3 Hybrid Combiners

The closer together the transmitter frequencies are on a T-Pass or Star combiner, the greater the insertion (combining) losses become. This is because the cavity filter coupling loops have to be adjusted so the filter has a narrow enough bandwidth to keep the neighboring (close) transmitter frequency out. There are limits then, given in Table 8.3, as to how close together the individual transmit frequencies can be on a T-Pass or Star combiner before the insertion losses become impractically high as a result.

If any of the transmitter frequencies that need to be combined are closer together than this, a hybrid combiner should be used instead. Hybrid

Table 8.3
Minimum T-Pass or Star Combiner Transmitter Frequency Separation

LMR Frequency Band	Minimum Transmitter Frequency Separation for T-Pass and Star Combiners
136–174 MHz	75 kHz
174–225 MHz	100 kHz
330–530 MHz	150 kHz
762–941 MHz	250 kHz

combiners are based on 3-dB, 90° hybrid couplers and can be used to combine any two uncorrelated signals with no fixed frequency or phase relationship.

To explain how a hybrid combiner works, let us start by considering a 3-dB hybrid coupler with one signal fed in, as shown in Figure 8.19.

Basically, the input signal (A) is split between the thru' (directly connected) port 4 and the coupled port 3. The signal at the thru' port 4 lags input by 90°, whereas the signal at coupled port 3 is in phase with the input.

Now let us consider what happens if we instead feed the signal in to port 2, as per Figure 8.20.

This time, the input signal (B) at port 2 is split between the thru' (directly connected) port 3 and the coupled port 4.

Now, if we consider Figure 8.21 where the two input signals A and B (i.e., the transmitters to be combined) are applied together, we will see half of each input signal at both output ports. What this means is that a 3-dB hybrid coupler can be used to combine two transmitters, but at a cost of half the power, as shown in Figure 8.22.

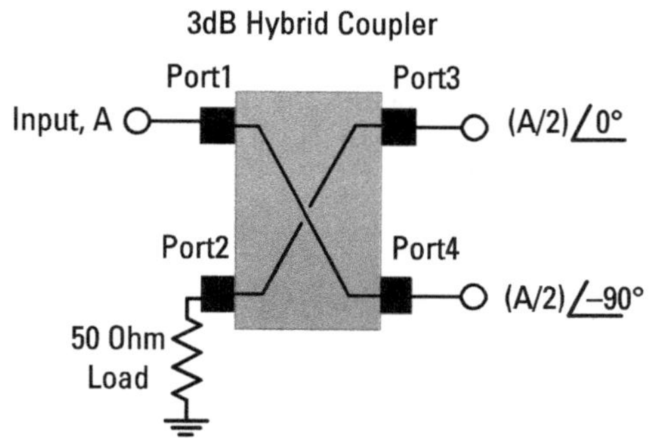

Figure 8.19 A 3-dB hybrid coupler with signal applied to port 1.

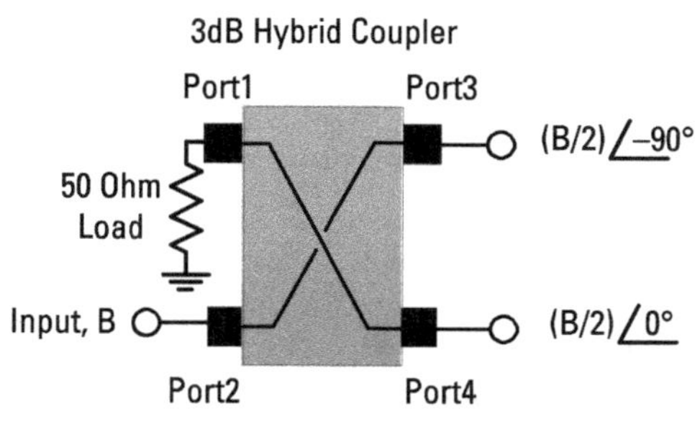

Figure 8.20 A 3-dB hybrid coupler with signal applied to port 2.

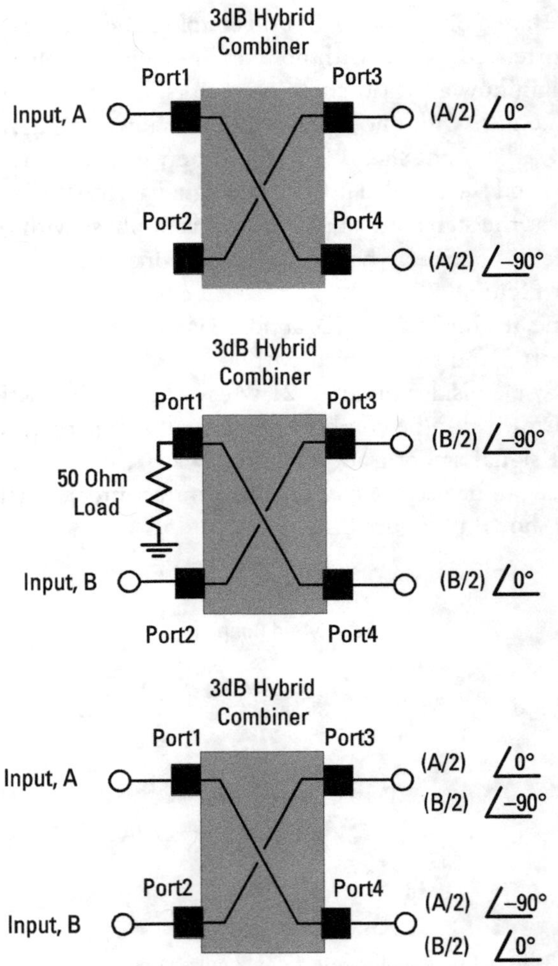

Figure 8.21 A 3-dB hybrid coupler with uncorrelated signals applied to ports 1 and 2.

Taking this further, Figure 8.23 shows how we would use multiple 3-dB hybrid couplers to combine more than two transmitters.

Unlike T-Pass or Star combiners, there is no limit on how close together the individual transmitter frequencies can be with hybrid combiners, and the combining losses are fixed no matter what the transmitter frequency separation is, as shown in Table 8.4.

Note that, in matched conditions (all ports terminated correctly in 50 ohms), the load must be capable of dissipating half the total power of the

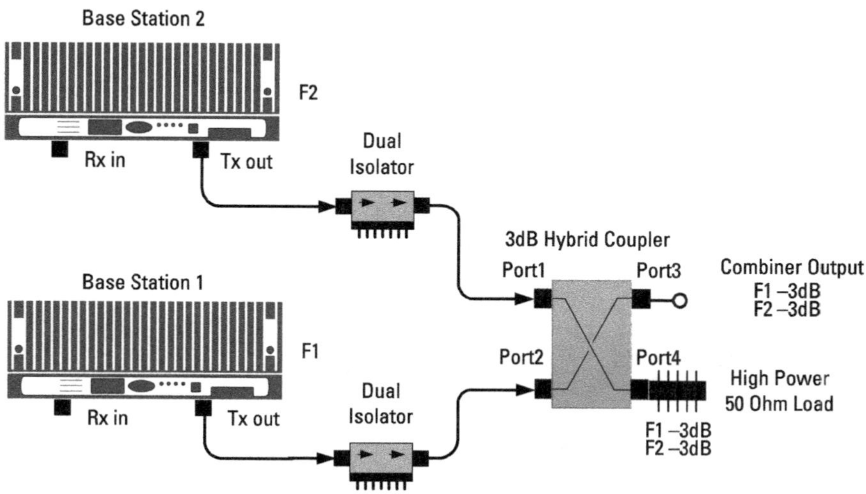

Figure 8.22 Two-channel hybrid combiner.

combined transmitters. However, because 3-dB hybrid couplers rely on each port being terminated in 50 ohms for correct operation, if the combined output is terminated in with a damaged cable, then the total power in the load can be even higher.

Unlike their cavity filter-based T-Pass and Star counterparts, hybrid combiners have no inherent filtering. Hence, hybrid combiners are almost always followed by a window filter at the output as per Figure 8.24, in other words, a filter that passes all the transmit frequencies but adequately suppresses transmitter sideband noise at the receive frequencies.

8.3.4 Combinations of Different Types

There are times when it makes sense to design a transmitter combining network consisting of more than one type of combiner. For example, to limit the

Table 8.4
Insertion Losses of Hybrid Combiners

Number of Combined Transmitters	Hybrid Combiner Insertion Loss
2	3.2 dB
4	6.4 dB
8	9.6 dB

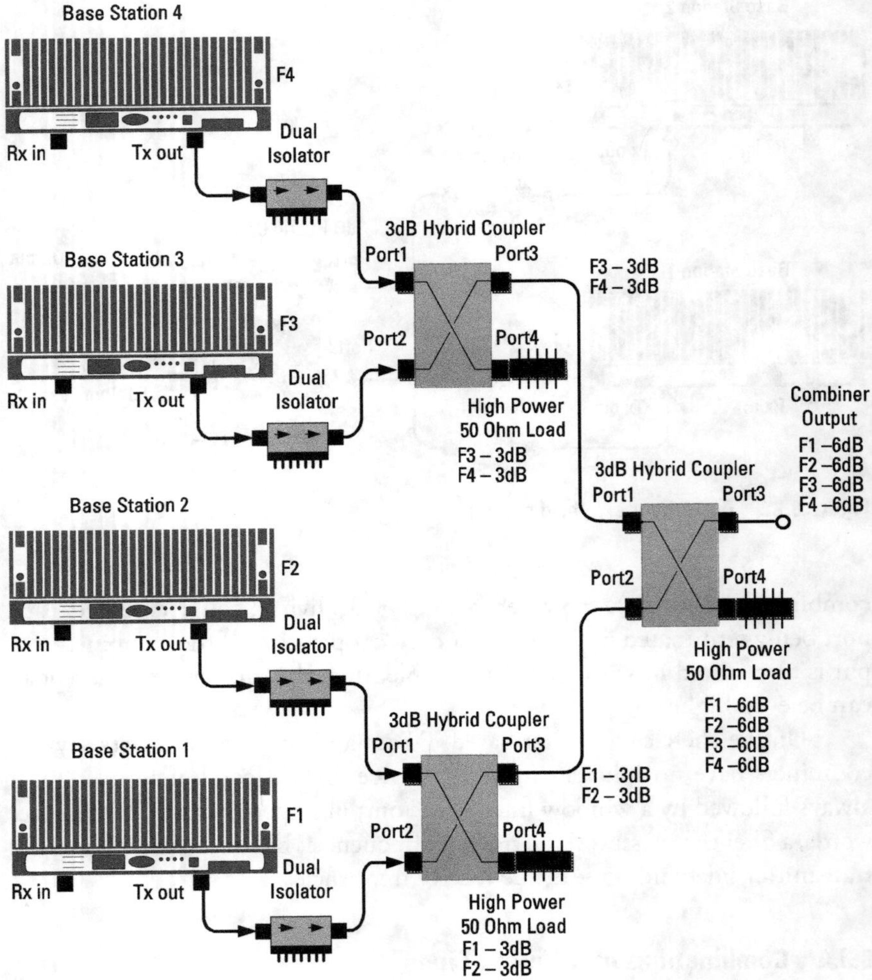

Figure 8.23 Four-channel hybrid combiner.

peak RF voltage that can be seen at the cavity filter outputs and minimize the difference in the individual insertion losses, we could design a four-channel combiner like that of Figure 8.25 where we T-pass-combine two groups of two channels and then Star-combine the outputs.

Equally, when combining two pairs of close-by transmitter frequencies with a suitable frequency interval between them, it makes sense to hybrid-combine the close pairs and then T-pass-combine the outputs as shown in Figure 8.26. Compared to a two-stage hybrid combiner, this arrangement will

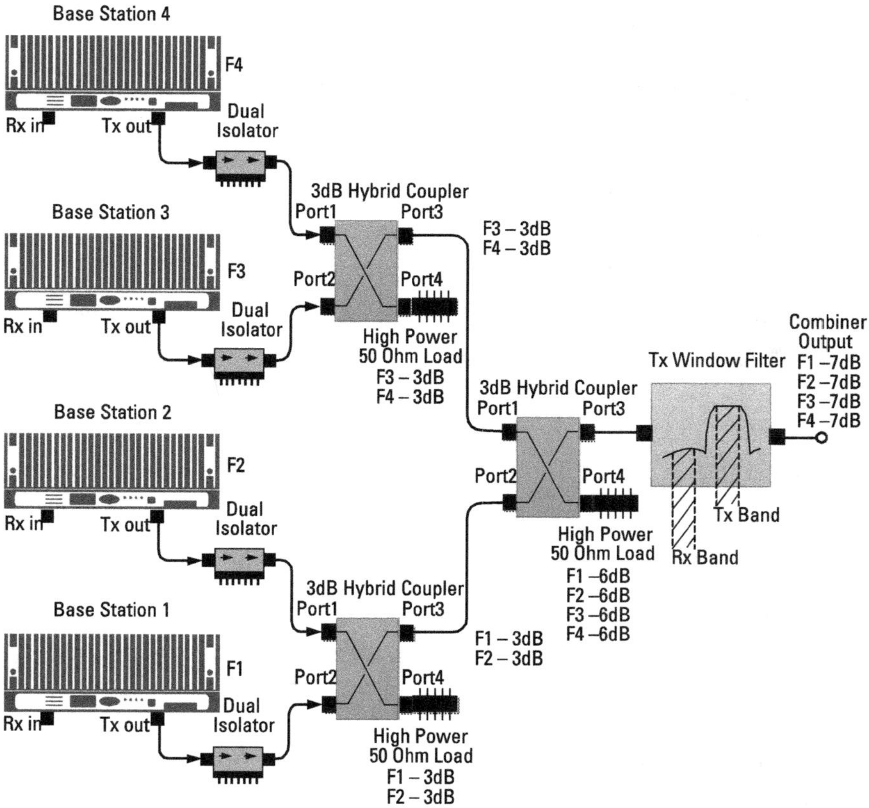

Figure 8.24 Four-channel hybrid combiner with Tx window filter.

typically have 1.5-dB lower insertion loss. Note that F_1 and F_2 must fit within the passband of cavity filter #1, and F_3 and F_4 must fit within the passband of cavity filter #2.

8.4 Specifying Transmitter Combiner Design Requirements

As well as combining several transmitters onto one antenna to save tower real estate, transmitter combiners perform two more important functions in managing interference:

- Mitigate transmitter IM.
- Clean up and suppress transmitter sideband noise at all site receive frequencies.

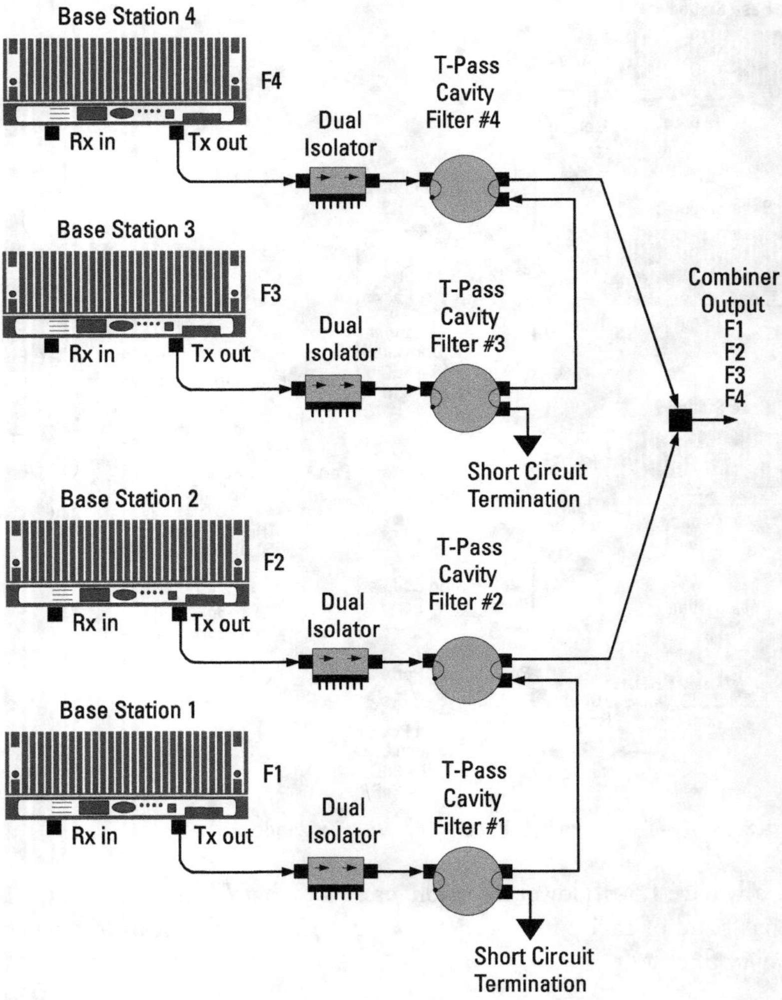

Figure 8.25 Four-channel T-Pass and Star combiners.

Both of these are vitally important so that we do not deafen any of the receivers at the site and limit their ability to hear weak uplink signals from users in the field.

Considering transmitter IM first (perhaps revisit the description in Chapter 4), for multichannel systems with separate transmit and receive antennas, the first thing we need to do is an IM analysis to determine if any IM products generated within the combiner by the mixing of the transmitter frequencies

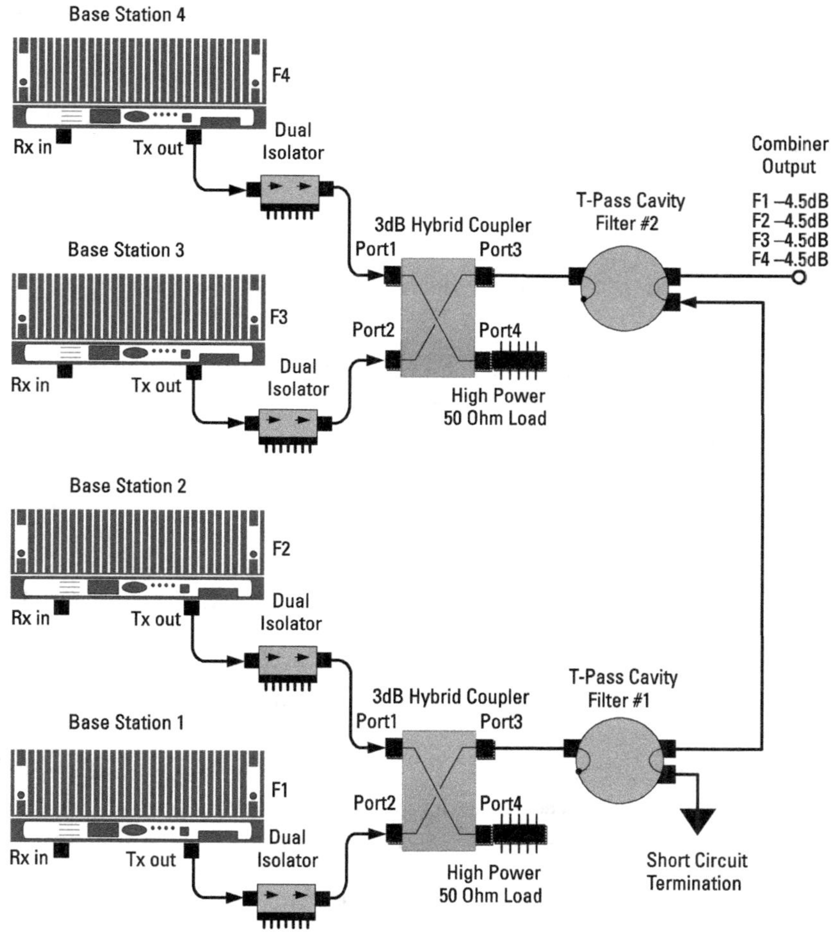

Figure 8.26 Four-channel hybrid and T-Pass combiners.

fall on any of the receive channels. For systems where we have 60 dB of isolation between the transmit and receive antennas, it is necessary to search for transmitter IM products out to seventh order within a search bandwidth of:

$$\text{Search bandwidth (kHz)} = (n + 1) \times \text{Channel bandwidth} \qquad (8.3)$$

where

n = maximum order of IM to be searched for;

Channel bandwidth = 12.5 kHz (typically for LMR systems).

For seventh-order IM and a 12.5-kHz channel bandwidth, the search band is therefore 100 kHz (+/–50 kHz around each receive frequency). This wider search bandwidth is necessary because the bandwidth of an IM product spreads according to its order, so while the center of the IM product may not fall directly on a receive channel, some of the energy can spill over into the receiver's passband.

Once the transmitter IM search has been carried out involving all the combined transmit frequencies, if IM products of seventh order or stronger (third, fifth, or seventh order) are found, then mitigation must be achieved by (in order of preference):

- Changing a transmit frequency so that the IM product no longer falls within the search bandwidth of a receive frequency. The IM product is still generated, just not at a frequency that interferes with any of the site receivers.
- Using multiple transmit antennas, each with their own combiner. Only combine transmitters that do not generate IM products on any site receive frequency.
- If neither of these are possible, then specify the filtering required in each arm of the transmitter combiner to reduce the level of the IM product below the design threshold at the affected site receive frequency, so no more than 1 dB of desensitization of the victim receiver results.
  - It sounds simple, but it is actually very difficult because much of the IM produced in a transmitter combiner is passive IM generated either on the antenna (after the combiner) or at the output of the isolators.
  - Passive IM does not conform to the rules presented in Chapter 3, so we cannot calculate the levels with any confidence.
  - Instead, the most practical approach to mitigating transmitter IM is to specify the isolation that must exist between the individual transmitter outputs on the combiner. A figure of 80 dB is recommended; this forces the transmitter combining equipment vendor to use dual isolators in the outputs of each transmitter.

Now, considering transmitter sideband noise, we need to calculate and specify the sideband noise suppression required in each arm of the transmitter combiner so no more than 1 dB of desensitization of any site receiver results. To do this, we need to know the output power of each transmitter at the transmit antenna port, their individual transmitter sideband noise levels at the nearest receive frequency, the isolation between the transmit and receive antennas, and also the interference threshold (the level below which the transmitter sideband noise must be suppressed at the receiver input so as not to result in any more than 1 dB of desensitization of any of the connected receivers).

As an example, let us consider a four-channel transmitter combiner where for each transmitter:

- Transmitter power (per channel at the transmit antenna), $P_{TX} = +45$ dBm, and comes from the sum of the base station transmitter output power (+50 dBm), the transmitter combiner loss (3 dB), RF cabling loss (2 dB).
- Transmitter sideband noise at nearest receive frequency, $Tx\ SBN_O =$ −150 dBc/Hz (assumed in this example to be the same, but it could be different).

Also, assume that the isolation between the transmit and receive antennas is 60 dB (including effect of the antenna gains), and at the receiving end:

- Receiver equivalent noise bandwidth, $B = 8$ kHz;
- Cascaded noise figure of the susceptible receiving system (including antenna cabling, filters, multicoupler, and base station receiver), $Rx\ NF = 8$ dB.

As per the example in Chapter 7, we start at the transmit end by calculating the total transmitter sideband noise radiated by the transmit antenna at the nearest receive frequency:

$$\begin{aligned} Tx\ SBN\ (\text{dBm/8 kHz}) &= Tx\ SBN_O + P_{\text{TX}} \\ &+ [10 \times \text{LOG}_{10}\ (B)] + [10 \times \text{LOG}_{10}\ (Tx)] \end{aligned} \tag{8.4a}$$

where

$Tx\ SBN =$ transmitter sideband noise that falls in the receiver's bandwidth (dBm/8 kHz);

$Tx\ SBN_O =$ normalized transmitter sideband noise at the receive frequency (dBc/Hz);

$P_{\text{TX}} =$ output power per transmitter at the transmit antenna input (dBm);

$B =$ filter pass bandwidth of the receiver (Hz);

$Tx =$ number of transmitters on the combiner.

So, in this example:

$$\begin{aligned} &Tx\ SBN \text{ (dBm/8 kHz)} \\ &= Tx\ SBN_O + P_{\mathrm{TX}} + [10 \times \mathrm{LOG}_{10}\ (B)] + [10 \times \mathrm{LOG}_{10}\ (Tx)] \\ &= -150 + 45 + [10 \times \mathrm{LOG}_{10}\ (8 \times 10^3)] + [10 \times \mathrm{LOG}_{10}\ (4)] \\ &= -60 \text{ dBm/8 kHz} \end{aligned} \tag{8.4b}$$

Now, at the receiver end, we calculate the receiver thermal noise floor, N_O:

$$N_O \text{ (dBm)} = -174 + [10 \times \mathrm{LOG}_{10}\ (B)] + Rx\ NF \tag{8.5a}$$

where

N_O = thermal noise power referred to the base of the receive antenna (dBm);

B = filter pass bandwidth of the receiver (Hz);

$Rx\ NF$ = noise figure of the susceptible receiving system (including antenna cabling, filters, multicoupler, and base station receiver) (dB).

$$\begin{aligned} N_O \text{ (dBm)} &= -174 + [10 \times \mathrm{LOG}_{10}\ (8 \times 10^3)] + 8 \\ &= -127 \text{ dBm/8 kHz} \end{aligned} \tag{8.5b}$$

At this point, before calculating the interference threshold, we would normally add the site noise. However, in this case, let us assume that there is no discernable site noise, as this situation will result in the highest required transmitter combiner sideband noise suppression (because the receiver noise floor is at its lowest possible value). Hence, we can say that:

$$N_X = 10 \times \mathrm{LOG}_{10}\ (N_T + I_T) \tag{8.6}$$

where

N_X = total noise and interference floor referred to the base of the receive antenna (dBm);

N_T = total noise floor referred to the base of the receive antenna (mW); note this is the same value as N_O in this example as we are assuming there is no additional site noise;

I_T = interference threshold, the total interference level referred to the base of the receive antenna (mW).

So, to find the maximum permissible level of transmitter sideband noise at the receive antenna (in other words, the interference threshold, I_T), the equation becomes:

$$I_T = 10 \times \text{LOG}_{10} (N_X - N_T) \tag{8.7a}$$

where

I_T = interference threshold, the maximum permissible level of transmitter sideband noise at the base of the receive antenna (dBm);

N_X = total noise and interference floor referred to the receive antenna (mW);

N_T = total noise floor referred to the base of the receive antenna (mW).

Again, remember the units of the noise levels must be converted to milliwatts first. So, going back to the calculation, if the level of transmitter sideband noise (I_T) that we can add to the −127-dBm thermal noise floor (N_O) causes the sum (N_X) to rise by 1 dB, we can say that:

$$\begin{aligned} N_X &= N_T + 1 \text{ dB} \\ &= -127 + 1 \\ &= -126 \text{ dBm} \end{aligned} \tag{8.7b}$$

So, converting to milliwatts:

$$\begin{aligned} N_X \text{ (mW)} &= 10^{(N_X \text{ (dBm)}/10)} \\ &= 10^{(-126/10)} \\ &= 2.51 \times 10^{-13} \text{ mW} \end{aligned} \tag{8.7c}$$

and

$$\begin{aligned} N_T \text{ (mW)} &= 10^{(N_T \text{ (dBm)}/10)} \\ &= 10^{(-127/10)} \\ &= 2 \times 10^{-13} \text{ mW} \end{aligned} \tag{8.7d}$$

Then, by substituting in these values, we find the maximum permissible level of transmitter sideband noise at the base of the receive antenna (I_T):

$$\begin{aligned} I_T &= 10 \times \text{LOG}_{10} (N_X - N_T) \\ &= 10 \times \text{LOG}_{10} [(2.51 \times 10^{-13}) - (2 \times 10^{-13})] \\ &= -132.87 \text{ dBm} \end{aligned} \tag{8.7e}$$

Again, as per the duplexer example in Chapter 7, we need to jump ahead of ourselves briefly and consider that the base station receivers connected to the receiver multicoupler are also exposed to the remnants of the transmitted signals that pass through the receiver filter. This could cause interference by means of receiver blocking, in addition to the remaining transmitter sideband noise that we have been considering up to now. The interference threshold must be modified to account for both mechanisms so that the total interference only results in 1-dB loss of receiver sensitivity. Revising from Chapter 6, if multiple interference mechanisms are present:

$$\text{Interference per mechanism (dBm), } I_T' = I_T - [10 \times \text{LOG}_{10}\ (n)] \quad (8.8a)$$

where

I_T = interference threshold, the total interference level at the base of the Rx antenna (dBm);

n = number of interference mechanisms simultaneously occurring.

So, in this case:

$$\begin{aligned}&\text{Interference per mechanism (dBm), } I_T' \\ &= I_T - [10 \times \text{LOG}_{10}\ (n)] \\ &= -132.87 - [10 \times \text{LOG}_{10}\ (2)] \\ &= -135.87 \text{ dBm}\end{aligned} \quad (8.8b)$$

We have now calculated the total transmitter sideband noise level on the transmit antenna at the nearest receive frequency (*Tx SBN*), and the maximum permissible level (I_T') that we can tolerate at the base of the receive antenna if we are only going to allow 1 dB of receiver desensitization when all transmitters are operating. So, the required sideband noise suppression in is simply the difference between the two, that is:

$$\begin{aligned}&\text{Transmitter sideband noise suppression at receive frequency (dB)} \\ &= Tx\ SBN - I_T \\ &= -60 - (-135.87) \\ &= 76 \text{ dB}\end{aligned} \quad (8.9)$$

This figure is the total transmitter sideband noise suppression required between the base of the transmit antenna and the base of the receive antenna. Of this, 60 dB is achieved simply by the isolation between the transmit and receive antennas, meaning, in this example, only the remaining 16 dB of transmitter sideband noise suppression is required from the cavity filters in each arm of the transmitter combiner.

To wrap up these calculations pictorially then, a frequency response plot of the example cavity filter for this channel in our transmitter combiner would look like Figure 8.27. Note that it is sensible to add an extra 5 dB to the calculated rejection figures to account for temperature drift and aging.

Finally, to give the transmitter sideband noise suppression calculation a little more meaning, it is better presented on a diagram as in Figure 8.28.

Specifying the performance that we need to, the chosen transmitter combiner vendor is best done by means of a table. For our example transmitter combiner, it would look like Table 8.5.

8.5 Differences for Duplexed Multichannel Systems

Despite strongly recommending the use of separate transmit and receive antennas for multichannel RF systems, sometimes there is no choice, and a single duplexed antenna has to be used. The reasons for this may be the lack of suitable locations on the tower for two antennas or the existing tower structural loading.

The design requirements for a transmitter combiner to be used within a duplexed multichannel system are specified in much the same way as before, with a couple of key differences. First, as explained in Chapter 7, the lack of any antenna isolation means that there is an increased risk of passive IM and

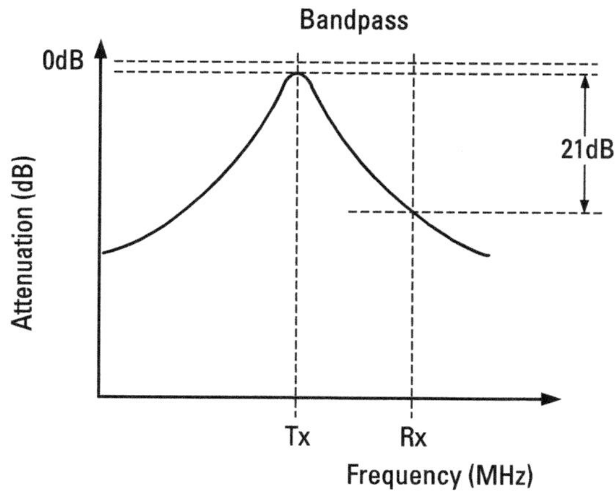

Figure 8.27 Transmitter combiner cavity filter frequency response plot.

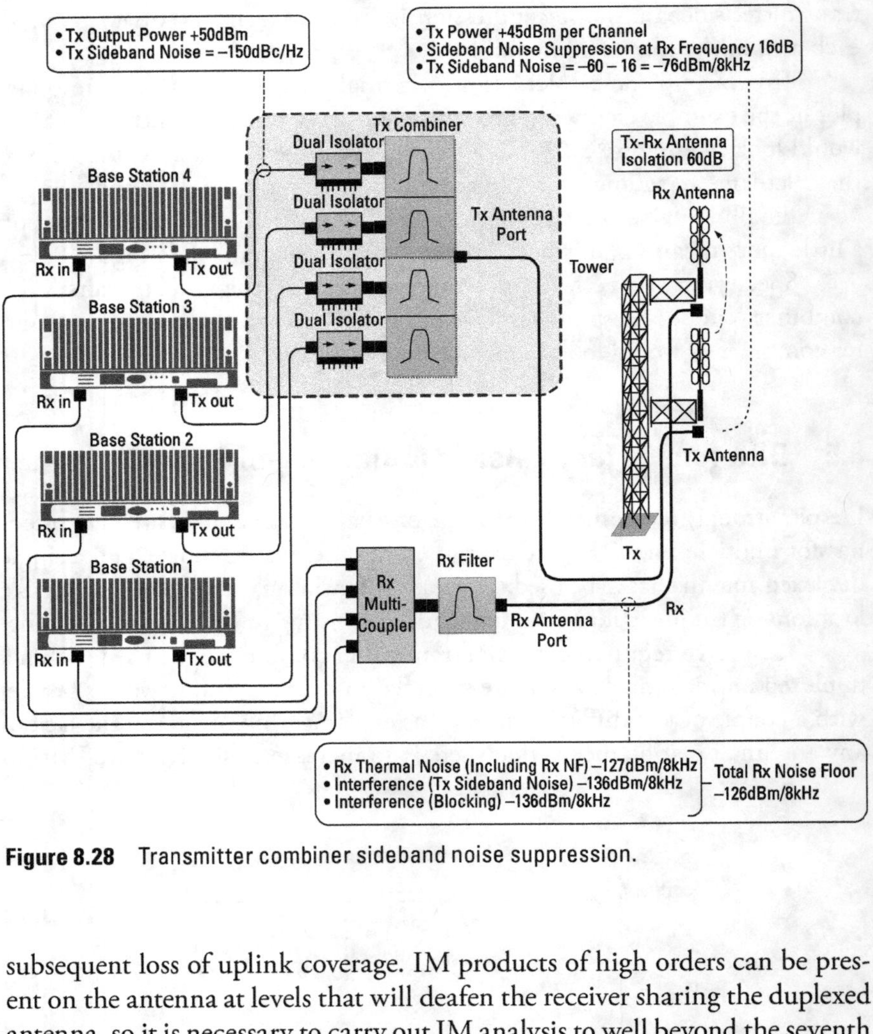

Figure 8.28 Transmitter combiner sideband noise suppression.

subsequent loss of uplink coverage. IM products of high orders can be present on the antenna at levels that will deafen the receiver sharing the duplexed antenna, so it is necessary to carry out IM analysis to well beyond the seventh order. The highest-order IM product that I have ever known to noticeably interfere with a duplexed receiver was the nineteenth-order result of three of the combined transmitter frequencies. It is therefore recommended that, for duplexed multichannel systems, transmitter IM searches be carried out to the twenty-first order, which makes frequency planning much harder.

Regarding transmitter sideband noise, all our sideband noise suppression must now be achieved in the transmitter combiner and the duplexer transmit to antenna filter. In the example calculation presented earlier in this chapter, the use of a duplexed antenna would mean all 76 dB of transmitter

Table 8.5
Example Transmitter Combiner Specification

Parameter	Design Limit
Number of transmit channels	4
Filter characteristic	Bandpass
Combiner type	T-Pass
Transmitter frequencies (the frequencies given here are simply examples)	F_1 = 160.225 MHz F_2 = 160.675 MHz F_3 = 161.450 MHz F_4 = 161.925 MHz
Insertion (combining) loss	≤4.0 dB
Tx port return loss in Tx passband	Better than 22 dB
Isolation between combined transmitter outputs	80 dB
RF input power per channel	≥100W (+50 dBm)
Offset to closest receive frequency (again, an example)	2 MHz
Sideband noise suppression at nearest receive frequency	≥21 dB
Input connectors (transmitter side)	Type N Female
Output connector (antenna side)	7/16DIN Female

sideband noise suppression would have to be achieved in the combiner hardware, which, in turn, means more insertion loss and thus less radiated power and downlink coverage.

8.6 Site Transmitter Combiner Design Rules

Before concluding with some best-practice transmitter combiner design rules, it is worth reminding ourselves of the points made in this chapter, that the proper design of the transmitter combiner is vital and gives following benefits:

- Maximum coverage from each connected transmitter by minimizing the combining losses;
- Minimum interference potential: Sideband noise, IM, and other unwanted emissions from the connected transmitters are all reduced below the design threshold and cause less than 1-dB desensitization of any site receiver.

The following are general best-practice design guidelines that describe how to design optimal transmitter combiners for deployment on radio towers.

1. Use separate transmit and receive antennas for multichannel systems: Do not use duplexers for multichannel RF system designs. Sometimes it is unavoidable due to tower congestion or loading, but, as a general rule, avoid multichannel duplexed RF systems.
2. Do not combine any more than four transmitters on to a single antenna, otherwise:
   - The RF power and voltage at the output loop of the top cavity filter of a T-Pass combiner may become dangerously high, especially in the event of a fault in the following antenna system.
   - The combining losses will become excessive.
   - The chances of generating IM on a site receive frequency start to increase dramatically.
3. The transmitter combiner insertion loss shall be no more than 4 dB (not including associated transmit antenna feeder cable losses). If the insertion loss exceeds this figure, consideration should be given to dividing the design into two transmitter combiners each feeding their own antenna.
4. Filtering in transmitter combiners must be predominantly bandpass:
   - Never use band-reject (notch) filtering alone in transmitter combiners.
   - A combination of bandpass and band-reject filtering is acceptable to provide extra sideband noise suppression at specific (usually very close) receive frequencies.
5. The minimum transmit to transmit frequency separation for T-Pass or Star combiners is shown in Table 8.6.

   If transmitters with frequency separations closer than this are to be combined, then either:
   - Use two transmitter combiners, each with their own antenna, and separate the transmitters accordingly.
   - Use a 3-dB hybrid combiner to combine the transmitters.
6. The transmitter combiner shall be designed to handle 100W of input power per connected transmitter.
   - Design T-Pass combiners so that the output loop of the top cavity filter is capable of handling the sum of the output powers of all the connected transmitters.
   - For hybrid combiners, use a load that is capable of handling the sum of the output powers of all the connected transmitters. In the event of the combined output being terminated in an open or short circuit, much more than half the reflected power ends up in the load.

Table 8.6
Minimum T-Pass or Star Combiner Transmitter Frequency Separation

LMR Frequency Band	Minimum Transmitter Frequency Separation
136–174 MHz	75 kHz
174–225 MHz	100 kHz
330–530 MHz	150 kHz
762–941 MHz	250 kHz

7. No more than 1 dB of desensitization of any site receiver shall occur due to IM generated in the transmitter combiner.
   - For systems with separate transmit and receive antennas that have 60 dB of isolation between them, transmitter IM analysis shall be carried out to seventh order within a search bandwidth of:

$$\text{Search bandwidth (kHz)} = (n + 1) \times \text{Channel bandwidth} \qquad (8.10)$$

where n = maximum order of IM to be searched for.

For seventh-order IM and a channel bandwidth of 12.5 kHz (typical for LMR systems), the search bandwidth is 100 kHz (+/–50 kHz either side of each receive frequency).

   - Once the transmitter IM search has been carried out involving all the combined transmit frequencies, if IM products of seventh order or stronger (third, fifth, or seventh order) are found within 50 kHz of any site receive frequency, then mitigation must be achieved by (in order of preference):
     - Changing a transmit frequency so that the IM product no longer falls within the search bandwidth of a receive frequency. The IM product is still generated at the same level, just not at a frequency that interferes with any of the site receivers.
     - Using multiple transmit antennas, each with their own transmitter combiner. Only combine transmitters that do not generate IM products on any site receive frequency.
     - If neither of these are possible, specify that 80 dB of isolation must exist between the individual transmitter outputs on the combiner (see design rule 8).
   - If a duplexed multichannel system with a common transmit and receive antenna is unavoidable, transmitter IM analysis shall be carried out to the twenty-first order within a search bandwidth of:

$$\text{Search bandwidth (kHz)} = (n + 1) \times \text{Channel bandwidth} \quad (8.11)$$

where n = maximum order of IM to be searched for.

For a twenty-first-order IM and a channel bandwidth of 12.5 kHz (typically for LMR systems), the search bandwidth is 275 kHz (+/−137.5 kHz on either side of each receive frequency).

- Once the transmitter IM search has been carried out involving all the combined transmit frequencies, if IM products of the twenty-first-order or stronger are found within 137.5 kHz of any site Rx frequency, then mitigation must be achieved by changing a transmit frequency so that the IM product no longer falls within the search bandwidth of a receive frequency.

8. The minimum isolation between the outputs of combined transmitters shall be at least 80 dB.
   - This figure includes the reverse isolation of the isolator and any additional attenuation provided by the cavity filters in neighboring arms of the transmitter combiner.
   - Use a dual isolator between the output of each transmitter and the associated transmitter combiner input port.
   - The isolator load closest to the transmitter combiner input port shall be rated at a minimum of 100W.
9. The filtering required in the output of each transmitter shall be specified such that no more than 1 dB of desensitization of any site receiver occurs as a result of transmitter sideband noise.
10. The transmitter combiner shall have type N Female connectors at each transmitter input port, and a 7/16 DIN Female connector at the combined output port (to minimize PIM)

Selected Bibliography

RFI Technology Solutions SPD4044-3115 Isolator/Circulator data sheet.

RFI Technology Solutions 380-550MHz Cavity Filters data sheet, https://www.rfi.com.au/core/media/media.nl?id=1930303&c=3389566&h=vfSlLGN2tquCYDbkD9TzpeEf5umcgaMz46GNOTv0y3BYyBmS.

Telecommunications Industry Association *Report TSB-88.1-F,* Wireless Communications Systems—Performance in Noise and Interference Limited Situations—Part 1: Recommended Methods for Technology Independent Narrowband Performance Modelling, April 2022.

9

Receiver Multicoupling Systems

A receiver multicoupler allows a number of base station receivers to be connected to a common receive antenna, hence further minimizing real estate and loading requirements on the tower. As will be explained during this chapter, this is done in a way that results in little or no loss of receiver sensitivity in order to maintain the uplink (user to site) coverage.

The term receiver multicoupler is usually applied just to the distribution system that amplifies and splits the received signal between the connected receivers. However, for the purposes of this book, we will define the receiver multicoupler as the receiver filter (also known as a preselector or roofing filter) and the distribution system, because, from a design point of view, they must be considered together. Figure 9.1 shows this and is a reminder of where the receiver multicoupler resides in a multichannel repeater site RF system.

The design of the receiver multicouplers is the last step in the RF system design for each radio repeater site in the network. The receiver multicoupler design can only be started after the site survey, tower mapping, and spectrum and noise monitoring have been done, subsequent frequency planning has been carried out, antenna placements on the tower have been finalized, and the receiver multicoupler design has been completed.

There are several types of receiver multicouplers, but, most commonly, this part of the system consists of a bandpass receiver filter (that passes all the desired receive signals and rejects unwanted interference), followed by a low

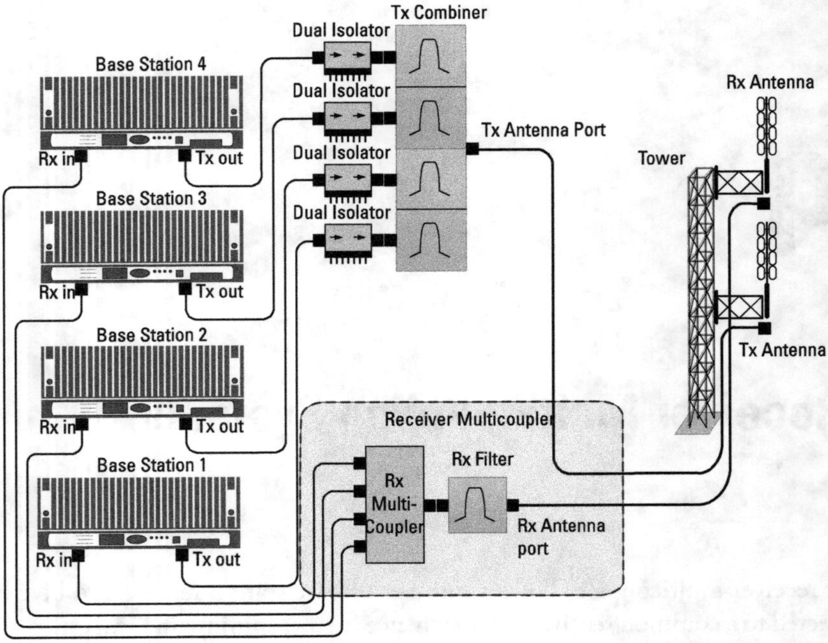

Figure 9.1 Receiver multicoupler in a multichannel site RF system.

noise amplifier (LNA), a variable attenuator, and a passive splitter. However, there are variations on this theme that sometimes have to be employed to prevent strong interfering signals from colocated transmitters at the site getting in and deafening the receivers.

In this chapter, you then will learn about the various types of receiver multicouplers and how they work so you can choose the correct type for each site in your system. We will then go into detail about how to calculate the receiver filter requirements. You do not have to physically design these yourself; instead, the aim of this chapter is to teach you enough to be able to select the correct model from your chosen vendor. Before that, though, it would be useful to quickly provide some background on receiver sensitivity, as this is the main system performance parameter that we are trying to maintain with the multicoupling system.

9.1 Receiver Sensitivity

For an LMR voice communications system, the sensitivity of a receiver is defined as the minimum RF input signal level from which the receiver can detect an intelligible audio output. This depends on the receiver's bandwidth, noise figure, and the minimum SNR required at the receiver demodulator (the detector where the audio is extracted from the RF carrier). Obviously, the better the receiver sensitivity is, the greater the coverage will be.

There are two ways that the receiver sensitivity can be expressed: static and dynamic. These were first covered in Chapter 6, but it is worth quickly revisiting them now.

Static sensitivity is the sensitivity of the receiver measured in a laboratory environment where the radio is directly connected to a signal generator. The RF input signal to the receiver is therefore constant in power level (contains no fading) and the only noise present is thermal noise resulting from the receiver's bandwidth and noise figure. Taking the example of a P25 Phase 1 site receiving system (the cascade of the multicoupler and following receivers) using C4FM modulation, the static sensitivity for DAQ3.0 would be calculated as follows by working out the receiver thermal noise floor and then adding the required static carrier-to-noise ratio (C_S / N) for that level of performance. Let us assume that the narrowest filter in the base station receiver has a pass bandwidth of 8 kHz, the cascaded noise figure of the site receiving system is 8 dB, and the required static carrier-to-noise ratio is 7.6 dB (from the TIA standard TSB-88.1-D).

$$N_O \text{ (dBm)} = -174 + [10 \times \text{LOG}_{10} (B)] + Rx\ NF \tag{9.1a}$$

where

N_O = thermal noise power referred to the base of the receive antenna (dBm);

B = filter pass bandwidth of the susceptible receiver (Hz);

$Rx\ NF$ = noise figure of the susceptible receiving system (including antenna cabling, filters, multicoupler and base station receiver) (dB).

So, for this example:

$$\begin{aligned} N_O \text{ (dBm)} &= -174 + [10 \times \text{LOG}_{10} (B)] + Rx\ NF \\ &= -174 + [10 \times \text{LOG}_{10} (8 \times 10^3)] + 8 \\ &= -174 + 39 + 8 \\ &= -127 \text{ dBm/8 kHz} \end{aligned} \tag{9.1b}$$

Then, using this to calculate the static sensitivity:

$$\text{Receiver static sensitivity (dBm)} = N_O + (C_S / N) \tag{9.2a}$$

where

N_O = thermal noise power referred to the base of the receive antenna (dBm);

C_S / N = the required constant level carrier-to-thermal noise ratio (dB), which, in this case, is 7.6 dB for DAQ3.0 using P25 C4FM modulation.

So

$$\begin{aligned} \text{Receiver static sensitivity (dBm)} &= -127 + 7.6 \\ &= -119.4 \text{ dBm} \end{aligned} \tag{9.2b}$$

Figure 9.2 shows the calculation in the form of a diagram.

Static sensitivity is a repeatable laboratory measurement, but it is only useful as a figure of merit that partially defines how good a receiver is. It is not the actual sensitivity that the site receiving system achieves in a real radio network, for two reasons:

- Noise at the radio site will be higher than thermal noise because of the other radio and electronic equipment installed there.
- The signal level picked up by the site receiver is not constant; it varies with time due to multipath fading as the transmitting radio terminal moves around.

Dynamic sensitivity takes account of these factors and is therefore the real sensitivity that the site receiving system achieves in practice. Taking the example of the same P25 Phase 1 receiving system using C4FM modulation,

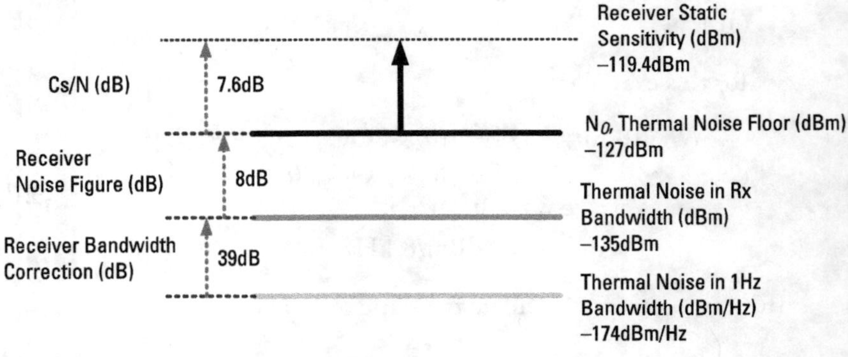

Figure 9.2 Receiver static sensitivity.

we can calculate the dynamic sensitivity for DAQ3.0 by working out the total noise floor and then adding the dynamic carrier-to-noise ratio ($C_F / (N + I)$) required for that level of performance. To calculate the total noise, we take the noise monitoring results from the RF survey (see Chapter 2) and look up the site noise level at the receive frequency in question. For this example, let us assume a site noise level of 10 dB above thermal was observed, which translated to an actual power level would be:

$$N_S \text{ (dBm)} = -174 + [10 \times \text{LOG}_{10} (B)] + S \tag{9.3a}$$

where

N_S = site noise referred to the base of the receive antenna (dBm);

B = filter pass bandwidth of the susceptible receiver (Hz);

S = site noise (dB above thermal noise).

So, for this example:

$$\begin{aligned} N_S \text{ (dBm)} &= -174 + [10 \times \text{LOG}_{10} (B)] + S \\ &= -174 + [10 \times \text{LOG}_{10} (8 \times 10^3)]] + 10 \\ &= -125 \text{ dBm} \end{aligned} \tag{9.3b}$$

Now that we know both the thermal noise (N_O) from our previous static sensitivity calculation and the site noise (N_S), the total noise (N_T) in the receiver's bandwidth can be found by simply adding the two together, that is:

$$N_T = 10 \times \text{LOG}_{10} (N_O + N_S) \tag{9.4a}$$

where

N_T = total noise referred to the base of the receive antenna (dBm);

N_S = site noise referred to the base of the receive antenna (mW);

N_O = thermal noise power referred to the base of the receive antenna (mW).

Note that the units, site noise (N_S) and thermal noise (N_O), are in milliwatts and the total noise floor (N_T) is in dBm. This is because site noise (N_S) and thermal noise (N_O) must be converted back to a linear quantity (milliwatts) first before they can be summed together.

$$\begin{aligned} N_O \text{ (mW)} &= 10^{(N_O \text{ (dBm)}/10)} \\ &= 10^{(-127/10)} \\ &= 1.99 \times 10^{-13} \text{ mW} \end{aligned} \tag{9.4b}$$

and

$$N_S\ (\text{mW}) = 10^{(N_S\ (\text{dBm})/10)} = 10^{(-125/10)} = 3.16 \times 10^{-13}\ \text{mW} \quad (9.4c)$$

So

$$N_T = 10 \times \text{LOG}_{10}\ (N_O + N_S) = 10 \times \text{LOG}_{10}\ [(1.99 \times 10^{-13}) + (3.16 \times 10^{-13})] = -122.9\ \text{dBm} \quad (9.4d)$$

To this we just need to add 1 dB, which we have previously defined in Chapter 6 as the acceptable deafening of the receiver as a result of external interference. So, our final noise and interference floor becomes:

$$N_X = N_T + 1 \quad (9.5a)$$

where

N_X = total noise and interference floor referred to the base of the receive antenna (dBm);

N_T = total noise floor referred to the base of the receive antenna (dBm).

So

$$N_X = N_T + 1 = -122.9 + 1 = -121.9\ \text{dBm} \quad (9.5b)$$

Now we can calculate the receiver dynamic sensitivity in the presence of thermal noise, site noise, and interference (N_X) by adding the relevant Rayleigh faded carrier-to-noise and interference ratio $C_F/(N+I)$. Compared to C_S/N, the difference is $C_F/(N+I)$ is a real-world threshold that includes the multipath variation in received signal level that happens because there is usually no line-of-sight path between the distant user who is transmitting and the receiver at the radio site.

As before, if we assume that our system is a P25 Phase 1 system using C4FM modulation and the customer requires a minimum DAQ of 3.0, then from Chapter 6, Table 6.1, the required $C_F/(N+I)$ is 15.2 dB. In which case, our final dynamic receiver sensitivity is:

$$\text{Receiver dynamic sensitivity (dBm)} = N_X + [C_F/(N+I)] = -121.9 + 15.2 = -106.7 \text{ dBm} \quad (9.6)$$

where

Receiver dynamic sensitivity = the received signal level required for minimum performance in the presence of thermal noise, site noise, and interference (dBm);

N_X = total noise and interference floor referred to the base of the receive antenna (dBm);

$C_F/(N+I)$ = the required Rayleigh faded carrier-to-noise and interference ratio (dB).

The important thing to note here is the real-world dynamic sensitivity of −106.7 dBm is nearly 13 dB worse than the static sensitivity of −119.5 dBm. When designing RF systems for shared radio sites, then we should always use dynamic sensitivity as that is what the receiving system is capable of in practice. Again, a diagram such as Figure 9.3 is the best way to illustrate the calculation.

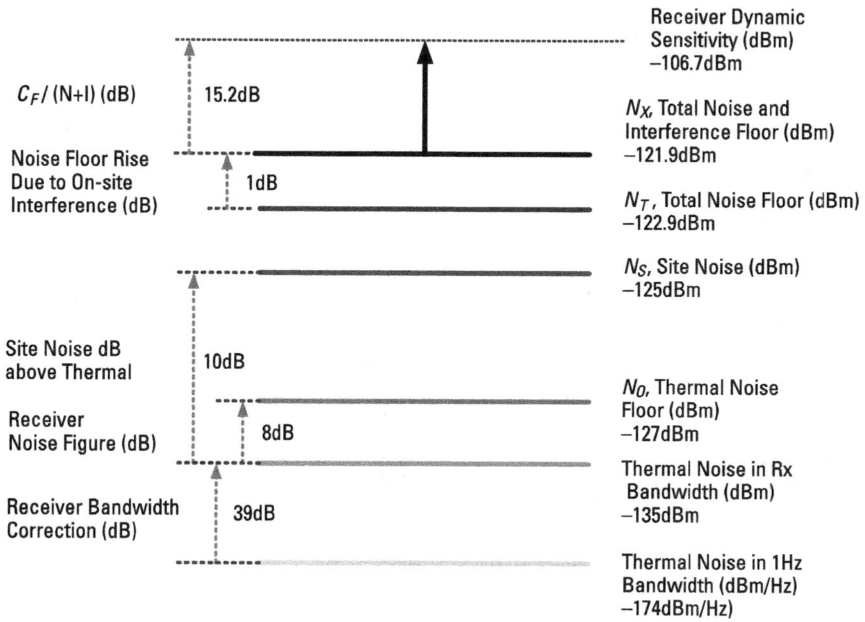

Figure 9.3 Receiver dynamic sensitivity.

9.2 Types of Receiver Multicouplers

There are two main types of receiver multicouplers: standard multicouplers and channelized multicouplers. Which one to use depends upon the relative frequencies of the connected receivers and the high-level interfering signals present at the site.

9.2.1 Standard Multicoupler

The standard multicoupler is the most basic type of multicoupling system, and consists of a receiver filter, LNA, variable attenuator, and passive splitter.

The standard multicoupler can be used where there are no high-level transmit signals at the site that lie within the passband of the required receiver (Rx) filter, as shown in Figure 9.4.

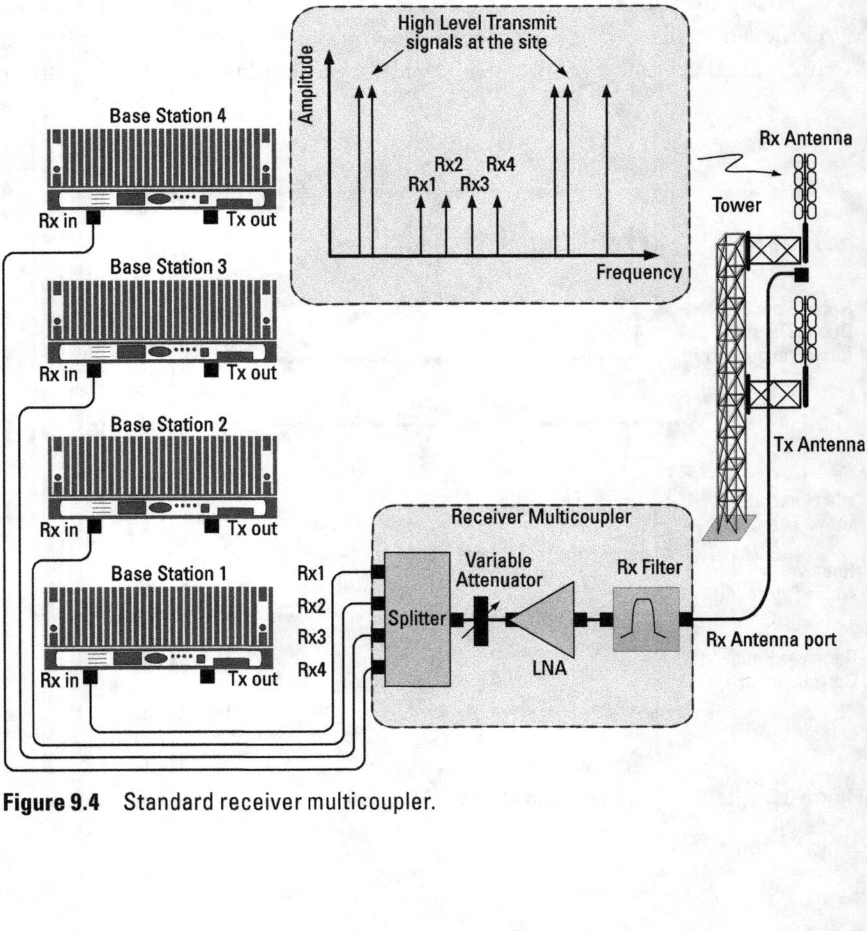

Figure 9.4 Standard receiver multicoupler.

Each of these parts has a specific role in the overall receiving system. The receiver filter is sometimes known as the preselector, roofing filter, or window filter. It is a narrow bandpass filter with steep rolloff that provides the rejection of high-level unwanted signals from all the transmitters at the site necessary to prevent any of the connected receivers being deafened due to blocking or IM. The fact that the transmit to receive frequency separation at UHF, for example, may be as little as 5 MHz means that most receiver filters are milled mechanical filters. At higher LMR frequencies, in the 800–900-MHz region, ceramic (dielectric) filters are sometimes used instead. Also, as the receiver filter comes before the LNA in the multicoupler lineup, it must have very low insertion loss; otherwise, the gain and noise figure requirements for the following LNA will become impractical.

Table 9.1 shows the typical performance of a UHF (400-MHz) receiver filter that would be used in a multicoupling system.

As mentioned at the beginning of this chapter, receiver multicouplers allow us to connect one receive antenna to multiple receivers. If we just split the signal between the connected receivers, we would progressively lower the signal level, degrade the sensitivity, and reduce the distance from which a user can be heard. So, we need an LNA in the receiver multicoupler that can overcome these splitting losses and maintain the sensitivity and IM (linearity) performance of the entire receiving system. To achieve this, the LNA must have:

- Enough gain and low enough noise figure to preserve the sensitivity performance of the entire receiving system;
- High enough third-order intercept point (IP3) to preserve the IM performance of the entire receiving system.

Table 9.1
Typical Receiver Filter Specifications

Parameter	**Performance**
Center frequency, F_C	400 MHz
Passband	F_C +/− 2 MHz
In-band insertion loss	<1.5 dB
In-band return loss	Better than 22 dB
Stopband	F_C +/− 7.5 MHz
Minimum rejection in stopband	>70 dB

Gain is simply the difference in level between the input signal level to the LNA and the resulting output signal level. The gain of LNAs used in receiver multicouplers is typically between 25 and 30 dB.

Noise figure (NF) was defined in Chapter 3 as the degradation of the SNR between the input and the output of the LNA. To maintain the sensitivity of the entire receiving system, the NF of the LNA must typically be better than 1.5 dB.

The IP3 was also defined in Chapter 3 and is a measure of the linearity of the LNA. The relevance of this is that unwanted interfering signals from site transmitters can mix in any nonlinearities in the receiver multicoupler and connected receivers to generate IM products that land at the same frequency as wanted receive signals. So, if the linearity of the LNA is not sufficient, the amplitude of any on-channel IM product may be high enough to deafen one or more of the connected receivers. To maintain the linearity of the entire receiving system, the output third-order intercept point (OIP3) of the LNA must be around +45 dBm.

Summarizing then, Table 9.2 shows the performance of an LNA suitable for use in a receiver multicoupler.

Receiver multicouplers also include a variable attenuator so that the gain, NF, and IP3 of the overall multicoupler can be adjusted correctly for the number of receivers that are connected. To avoid unduly affecting the NF (and hence sensitivity) of the overall receiving system, the attenuator is always installed after the LNA. The typical range of adjustment of the attenuator in a receiver multicoupler would be 0 to 15 dB in 1-dB steps.

The final part of a standard receiver multicoupler is the splitter, which simply divides the incoming received signal equally between the connected receivers. As the splitter is dividing the incoming signal, it obviously introduces loss, with each split reducing the signal level by 3 dB plus internal losses that amount to about 0.2 dB per split.

Table 9.3 gives the splitter loss as a function of the number of outputs.

Table 9.2
Typical Receiver Multicoupler LNA Specifications

Parameter	Performance
Gain	27 dB
NF	<1.5 dB
OIP3	>45 dBm

Table 9.3
Passive Splitter Losses

Number of Outputs	Number of Splits	Insertion Loss
2	1	3.2 dB
4	2	6.4 dB
8	3	9.6 dB
16	4	12.8 dB
32	5	16.0 dB

Receiver multicouplers need to be easily expandable so more channels can be added at a later date if necessary. This is achieved using a pyramid style of interconnection as illustrated in the following, starting with a simple eight-channel system. Eight-channel multicouplers with an expansion port, as shown in Figure 9.5, are what most vendors offer as a minimum.

Unlike transmitter combiners where there are limits as to how many transmitters it is sensible to combine on a single antenna (see design rule 2 in Chapter 8 for the reasons why), there are no such restrictions for receiver multicouplers. If we needed to expand the multicoupler to feed eight more receivers, an eight-way expansion splitter is simply connected to the expansion port, as shown in Figure 9.6.

Keeping this style of expansion ensures the signal level at each output remains the same (i.e., all the connected receivers see the same received signal level).

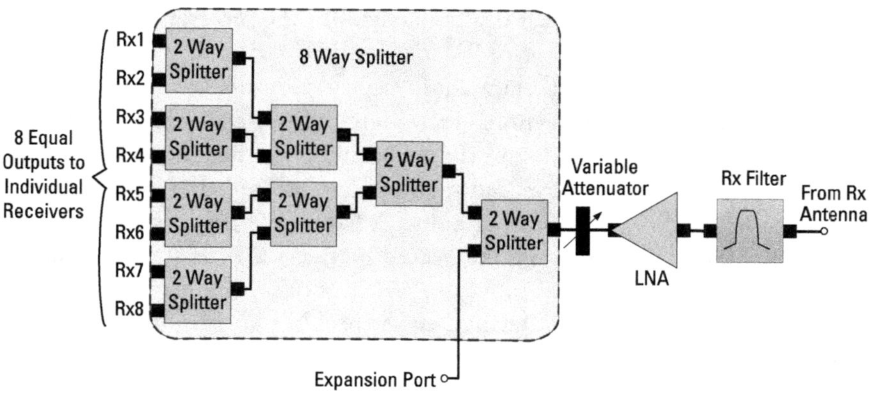

Figure 9.5 An 8-channel standard multicoupler with expansion port.

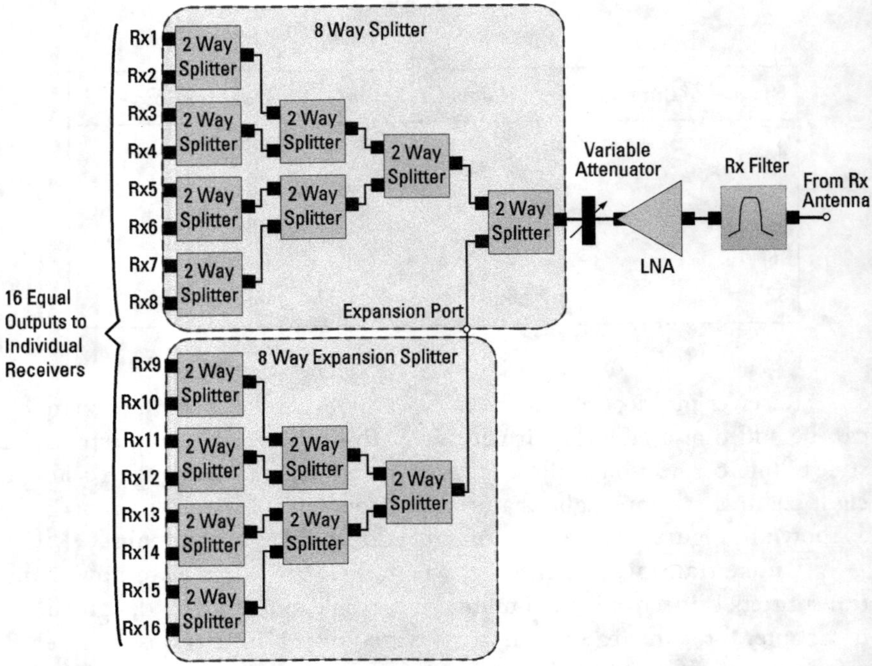

Figure 9.6 A 16-channel standard receiver multicoupler.

9.2.2 Channelized Multicoupler

There will be times when the frequency of one or more high-level transmitters at the site will fall in between the frequencies of the receivers connected to the multicoupler. In other words, if a standard multicoupler is used with one common receive filter, it will allow those high-level transmit carriers through as well. Deafening of one or more of the connected receivers due to blocking or IM would probably result, and the consequence of that would be a reduction in the distance that a user can maintain communication with the site.

So, when the frequencies of colocated transmitters at the site fall in between the frequencies of the connected receivers like this, a channelized multicoupler like that shown in Figure 9.7 must be used.

The difference is the filtering before the LNA. Figure 9.8 shows that instead of all the receive frequencies passing through one common receive filter, each receive frequency now has its own filter meaning the high-level transmit

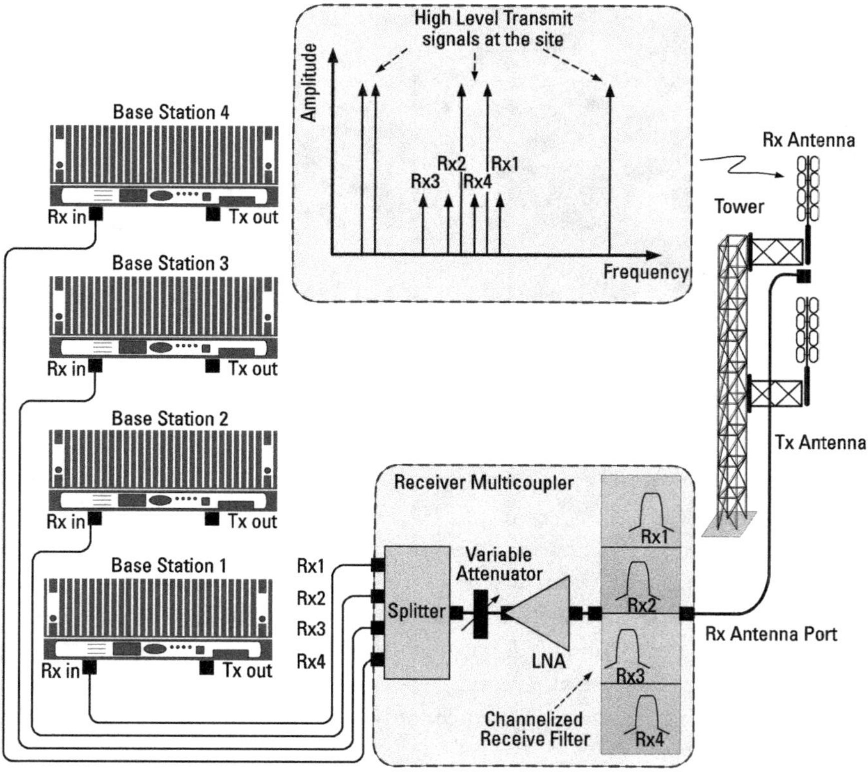

Figure 9.7 Channelized receiver multicoupler.

signals that lie in between are rejected. The channelized receive filter is likely to consist of highly selective cavity filters similar to those used in transmitter combiners, often in series with a milled receive window filter similar to that used in the standard multicoupler. The cavity filters provide the selectivity required to suppress the high-level transmitter carriers, while the milled window filter is still required because the cavity filters only have an ultimate rejection of around 35 dB (as opposed to at least 80 dB for a milled filter).

In the channelized multicoupler shown in Figure 9.8, a Star junction was used to carry out the initial splitting of the received signals. However, just as in transmitter combiners, either Star or T-Pass topologies can be used in receiver multicouplers to split the received signal and recombine afterwards. Figure 9.9 shows a channelized multicoupler using T-Pass receiver filtering.

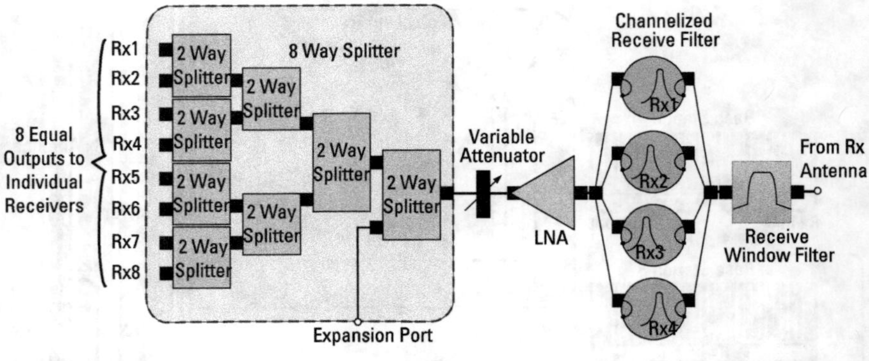

Figure 9.8 An 8-way channelized multicoupler.

A variation of the channelized multicoupler that achieves superior performance involves having separate LNAs and variable attenuators for each individual receiver. This configuration, shown in Figure 9.10, costs a little more but is especially useful when the frequency of one or more high-level transmitters is within a few hundred kilohertz of a receive frequency.

Basically, as explained in Chapter 8, Star connections are a compromise; the Star cable lengths are an average of the quarter-wavelengths of the tuned cavity filter center frequencies. Hence, only having a Star connection on the

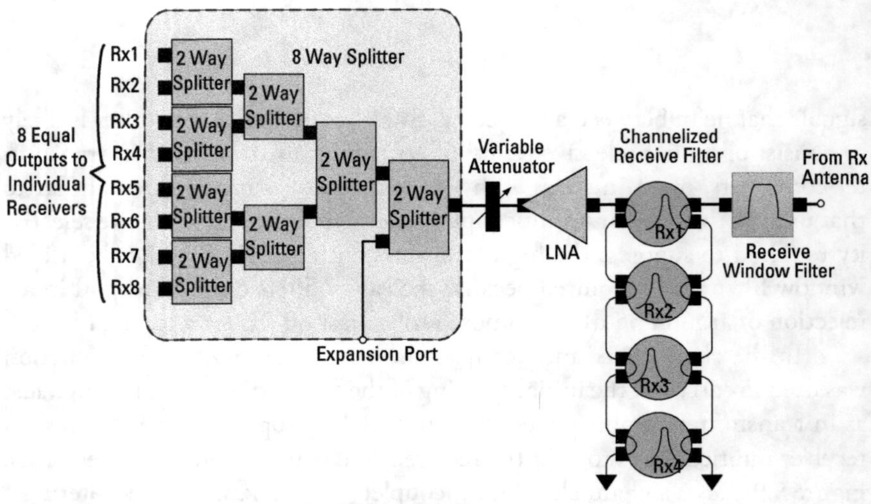

Figure 9.9 An 8-way T-Pass channelized multicoupler.

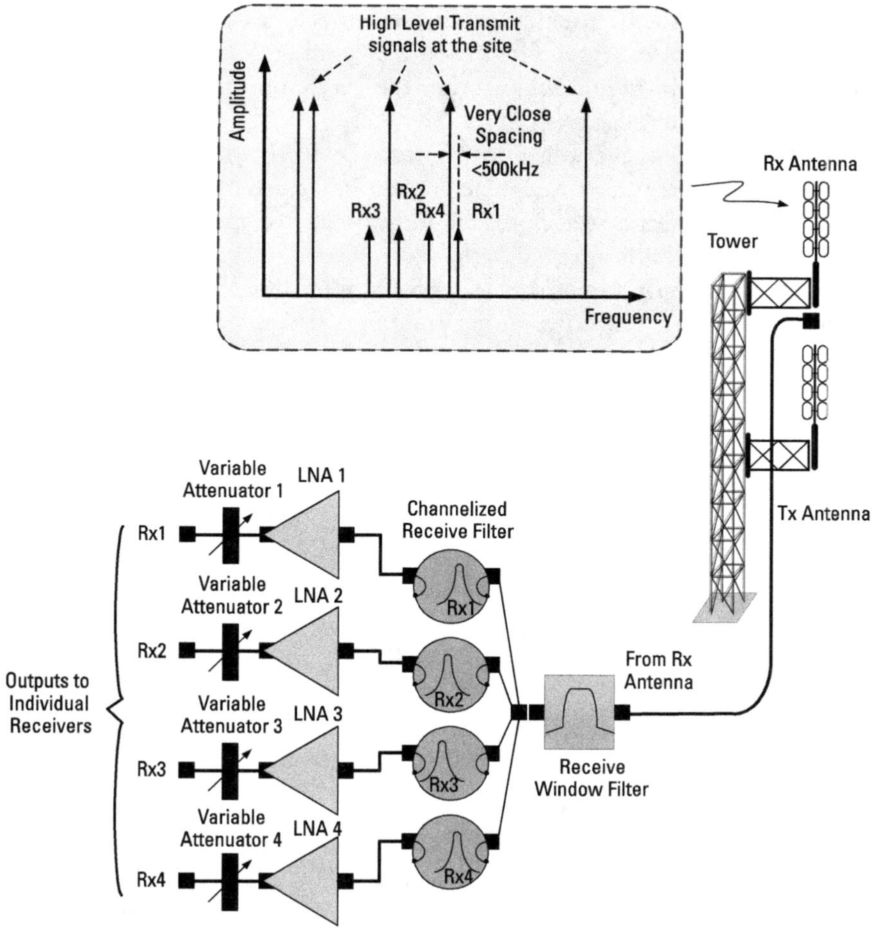

Figure 9.10 A 4-way channelized multicoupler with separate LNA and attenuators.

input of the cavity filters compromises the selectivity of the filter less than having the Star connection on the input and output would. The advantage is that the cavity filters have a more selective response and can therefore achieve a greater rejection of very close high-level carriers compared to the channelized multicoupler shown previously in Figure 9.8.

9.2.3 Use of Crystal Filters in Channelized Multicouplers

If a high-level transmit carrier is picked up by the receive antenna even closer to a receive frequency, say within 150 kHz, then the preceding cavity filters may not offer enough attenuation to prevent the deafening of the victim receiver. In this case, crystal filters should be used in that individual receiver input as shown in Figure 9.11 so that the very close transmit carriers are adequately suppressed.

This specific scenario is most common in VHF systems in the United States because there are no designated transmit or receive subbands. Transmit and receive channels are therefore randomly distributed and often intermingle, even at the same radio site.

The maximum input level to a crystal filter should be limited to −10 dBm; otherwise, it will suffer damage and/or generate IM of its own.

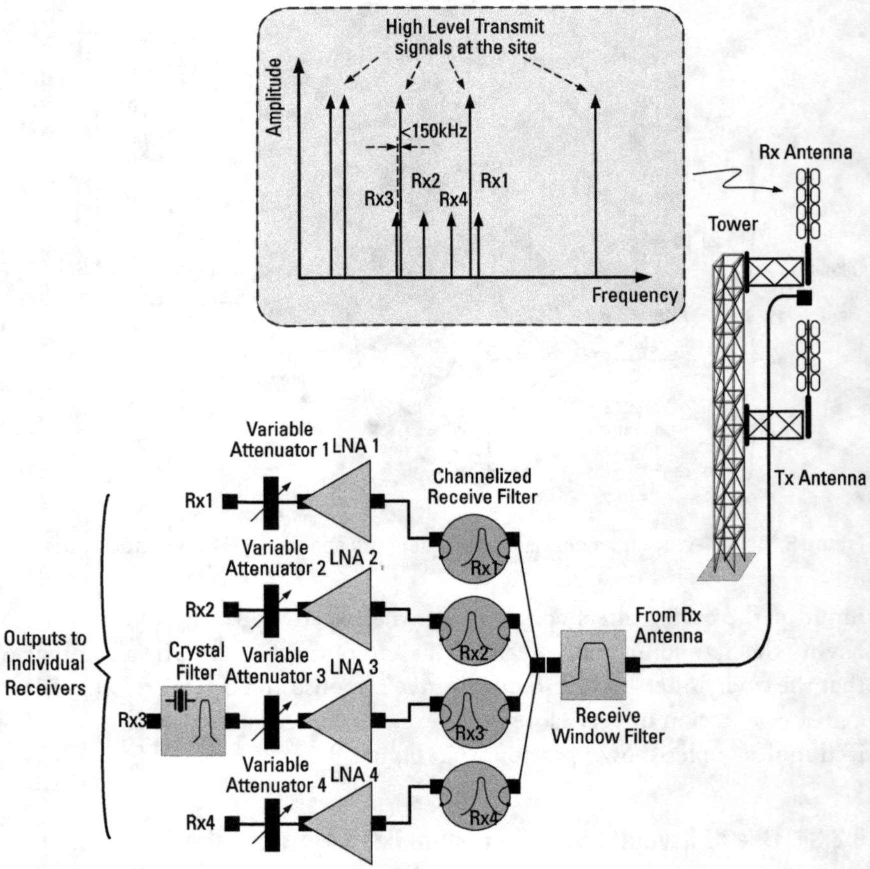

Figure 9.11 Four-way channelized multicoupler with a crystal filter.

9.3 Use of Tower Top Amplifiers in Receiver Multicouplers

All the types of receiver multicouplers through which we have worked so far are designed to provide mobile uplink coverage; in other words, the distance from which a user with a vehicle mounted mobile radio can communicate with the site will be roughly the same as the distance from which the site can reach that user. Sometimes, though, it is necessary to provide uplink coverage from users with handheld portable radios that typically transmit at a much lower power level (less than one-tenth) than mobile radios would. If nothing is done at the radio site to compensate for the lower transmit power of the portable radios, then the uplink coverage will be much less than that of the downlink, and the users with a portable radio would be able to hear the site from a much greater distance than they could reach it to establish a call; obviously not a good situation. The best thing to do about this is to improve the sensitivity of the site receiving system so it can pick up the weaker signals from the portable radios in the field. This is done using tower top amplifiers (TTAs), which simply consist of another bandpass filter and LNA mounted directly after the site receive antenna at the top of the tower. The gain of a TTA is typically around 20 dB, with an NF of 3 dB and OIP3 of +40 dBm.

To understand how a TTA gives sensitivity benefit, it is necessary to quickly go back and revisit NF from Chapter 3. NF is a figure of merit that quantifies the degradation in SNR that occurs when a signal is passed through an electronic device or system. In radio systems, it is most usually used to quantify the degradation in SNR that occurs when a wanted signal is passed through a receiver.

$$\text{NF (dB)} = \text{SNR at system input (dB)} - \text{SNR at system output (dB)} \quad (9.7)$$

So, in the case of our site receiving system (multicoupler plus connected base station receivers), while each part has its own individual NF, it is the overall NF of the whole lineup that really matters, as this is what determines the minimum RF signal level that the site receiving system can intelligibly detect.

To calculate the overall (termed "cascaded") NF of the site receiving system, first convert the gains and NFs of all the individual stages from decibels to pure ratios. Then, to find the overall NF as a pure ratio, use the following formula:

$$\begin{aligned}\text{Cascade NF, } F = F_1 + [(F_2 - 1)/G_1] + [(F_3 - 1)/(G_1 \times G_2)] \\ + [(F_4 - 1)/(G_1 \times G_2 \times G_3)] \ldots\end{aligned} \quad (9.8a)$$

where (all quantities are pure ratios with no unit):

F_1 = NF of the first part of the site receiving system (nearest the receive antenna);
G_1 = gain of the first part of the site receiving system;
F_2 = NF of the second part of the site receiving system;
G_2 = gain of the second part of the site receiving system;
F_3 = NF of the third part of the site receiving system;
G_3 = gain of the third part of the site receiving system;
F_4 = NF of the fourth part of the site receiving system.

Applying this to a simple four-channel site receiving system shown in Figure 9.12, we get:

$$\begin{aligned}
&\text{Cascade NF, } F = F_1 + [(F_2 - 1)/G_1] + [(F_3 - 1)/(G_1 \times G_2)] \\
&+ [(F_4 - 1)/(G_1 \times G_2 \times G_3)] + [(F_5 - 1)/(G_1 \times G_2 \times G_3 \times G_4)] \\
&+ [(F_6 - 1)/(G_1 \times G_2 \times G_3 \times G_4 \times G_5)] \\
&= 1.585 + [(1.412 - 1)/0.631] + [(1.412 - 1)/(0.631 \times 0.708)] \\
&+ [(10 - 1)/(0.631 \times 0.708 \times 501)] \\
&+ [(9.12 - 1)/(0.631 \times 0.708 \times 501 \times 0.1)] \\
&+ [(6.31 - 1)/(0.631 \times 0.708 \times 501 \times 0.1 \times 0.109)] \\
&= 1.585 + 0.653 + 0.922 + 0.040 + 0.362 + 2.177 \\
&= 5.739
\end{aligned} \tag{9.8b}$$

Then convert the cascaded NF to decibels, that is:

$$\text{NF (dB)} = 10 \times \text{LOG}_{10}\ (F) \tag{9.9a}$$

where F = cascade NF as a ratio.

So

$$\begin{aligned}
\text{NF (dB)} &= 10 \times \text{LOG}_{10}\ (F) \\
&= 10 \times \text{LOG}_{10}\ (5.379) \\
&= 7.59 \text{ dB}
\end{aligned} \tag{9.9b}$$

Note that the answer of 7.59 dB is similar to the 8-dB NF of the base station receivers alone. This is deliberate and neatly illustrates the whole point of receiver multicoupler: to allow one receive antenna to feed multiple receivers with no loss of sensitivity. In other words, the sensitivity of each connected

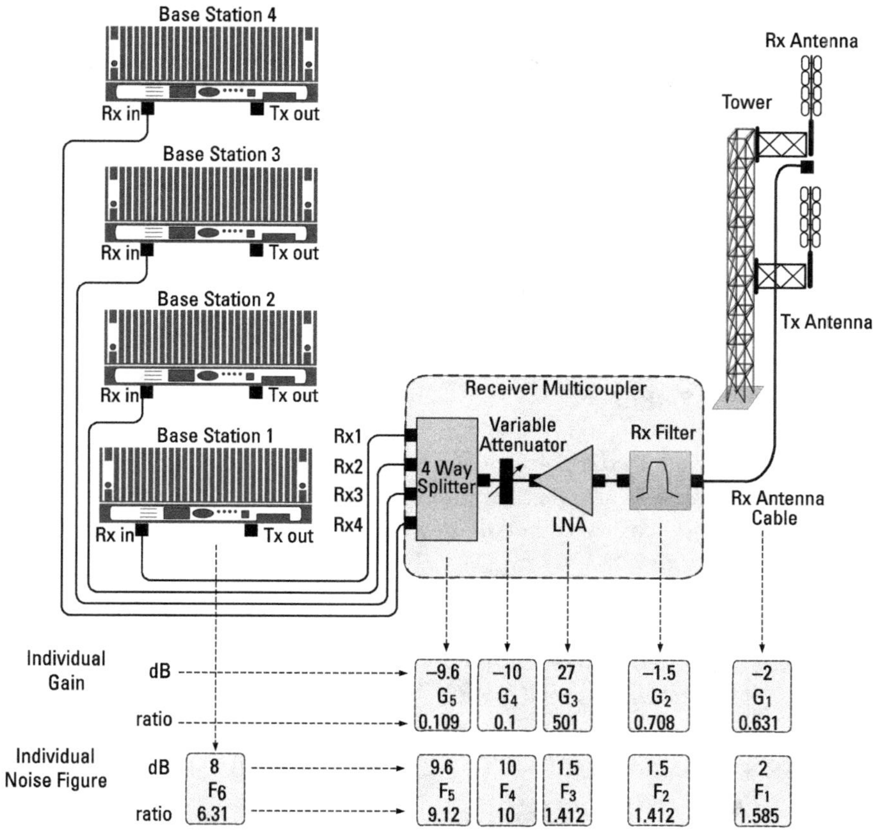

Figure 9.12 Site receiving system showing individual stage gains and noise figures.

receiver is the same as it would be if each receiver was directly connected to the base of its own receive antenna.

If we again assume our system is a P25 Phase 1 system using C4FM modulation and the customer requires a minimum DAQ of 3.0, then from Chapter 6, Table 6.1, the required $C_F/(N+I)$ is 15.2 dB, in which case, the dynamic sensitivity that this receiving system is capable of achieving without the addition of any site noise would be given by:

$$\text{Receiver dynamic sensitivity (dBm)} = -174 + [10 \times \text{LOG}_{10}(B)] + Rx\ NF + (C_F/N+I) \quad (9.10a)$$

where

B = filter pass bandwidth of the susceptible receiver (Hz);

$Rx\ NF$ = NF of the receiving system (including antenna cabling, filters, multicoupler, and base station receiver) (dB);

$C_F/(N+I)$ = the Rayleigh faded carrier-to-noise and interference ratio (dB).

Doing the calculation:

$$\begin{aligned}&\text{Receiver dynamic sensitivity (dBm)}\\&= -174 + [10 \times \text{LOG}_{10}\ (B)] + Rx\ NF + (C_F/N+I)\\&= -174 + [10 \times \text{LOG}_{10}\ (8 \times 10^3)] + 7.59 + 15.2\\&= -112.2 \text{ dBm}\end{aligned} \tag{9.10b}$$

Getting back to the point then, what happens if we place a TTA in the system, at the base of the receive antenna? Let us find out by repeating the NF and dynamic sensitivity calculations, but this time with the TTA added, as shown in Figure 9.13.

$$\begin{aligned}&\text{Cascade NF, } F = F_1 + [(F_2 - 1)/G_1] + [(F_3 - 1)/(G_1 \times G_2)]\\&+ [(F_4 - 1)/(G_1 \times G_2 \times G_3)] + [(F_5 - 1)/(G_1 \times G_2 \times G_3 \times G_4)]\\&+ [(F_6 - 1)/(G_1 \times G_2 \times G_3 \times G_4 \times G_5)]\\&+ [(F_7 - 1)/(G_1 \times G_2 \times G_3 \times G_4 \times G_5 \times G_6)]\\&= 2 + [(1.585 - 1)/100] + [(1.412 - 1)/(100 \times 0.631)]\\&+ [(1.412 - 1)/(100 \times 0.631 \times 0.708)]\\&+ [(10 - 1)/(100 \times 0.631 \times 0.708 \times 501)]\\&+ [(9.12 - 1)/(100 \times 0.631 \times 0.708 \times 501 \times 0.1)]\\&+ [(6.31 - 1)/(100 \times 0.631 \times 0.708 \times 501 \times 0.1 \times 0.109)]\\&= 2 + 0.0059 + 0.0065 + 0.0092 + 0.0004 + 0.0036 + 0.0218\\&= 2.0474\end{aligned} \tag{9.11}$$

Then convert the cascaded NF to decibels, that is:

$$\text{NF (dB)} = 10 \times \text{LOG}_{10}\ (F) \tag{9.12a}$$

where F = cascade NF as a ratio.

So

$$\begin{aligned}\text{NF (dB)} &= 10 \times \text{LOG}_{10}\ (F)\\&= 10 \times \text{LOG}_{10}\ (2.0474)\\&= 3.11 \text{ dB}\end{aligned} \tag{9.12b}$$

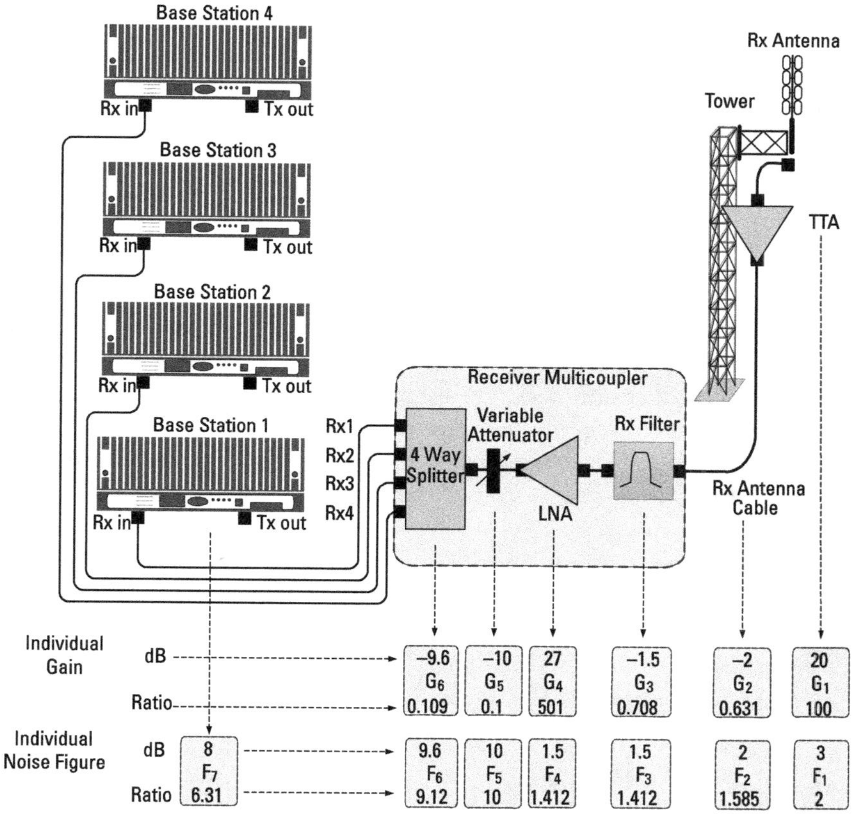

Figure 9.13 Cascaded noise figure of site receiving system with TTA.

With the addition of the TTA, the NF of our system has improved from 7.59 dB to 3.11 dB, an improvement of 4.48 dB.

Once more taking a P25 Phase 1 system using C4FM modulation, designed for a minimum DAQ of 3.0, the dynamic sensitivity this receiving system is now capable of achieving without the addition of any site noise would be given by:

$$\text{Receiver dynamic sensitivity (dBm)} = -174 + [10 \times \text{LOG}_{10}\,(B)] + Rx\,NF + (C_F\,/N + I) \tag{9.13a}$$

where

B = filter pass bandwidth of the susceptible receiver (Hz);

$Rx\ NF$ = NF of the receiving system (including antenna cabling, filters, multicoupler, and base station receiver) (dB);

$C_F/(N+I)$ = the Rayleigh faded carrier-to-noise and interference ratio (dB).

Doing the calculation:

$$\begin{aligned} &\text{Receiver dynamic sensitivity (dBm)} \\ &= -174 + [10 \times \text{LOG}_{10}\ (B)] + Rx\ NF + (C_F/N+I) \\ &= -174 + [10 \times \text{LOG}_{10}\ (8{,}000)] + 3.11 + 15.2 \\ &= -116.7\ \text{dBm} \end{aligned} \tag{9.13b}$$

This is 4.5 dB lower than the system achieved without the TTA. So, placing a TTA before the losses of the main receive antenna transmission line and the receiver filter means a lower overall site receiving system noise figure is possible. The site receiving system can then adequately detect weaker signals, meaning that uplink coverage is extended, that is, the user with a handheld portable radio can now make contact with the site from a much greater distance than before.

If something sounds too good to be true, then it might be too good to be true. In this case, it depends on the ambient site noise, and it is best illustrated by repeating the sensitivity calculations with and without the TTA, but with the addition of site noise 10 dB above thermal.

First, without the TTA, we start by calculating the total noise, N_T, from the thermal noise, N_O, and the site noise, N_S:

$$N_O\ (\text{dBm}) = -174 + [10 \times \text{LOG}_{10}\ (B)] + Rx\ NF \tag{9.14a}$$

where

N_O = thermal noise power referred to the base of the receive antenna (dBm);

B = filter pass bandwidth of the susceptible receiver (Hz);

$Rx\ NF$ = NF of the receiving system (including antenna cabling, filters, multicoupler, and base station receiver) (dB).

So, for this example, the thermal noise is:

$$\begin{aligned} N_O\ (\text{dBm}) &= -174 + [10 \times \text{LOG}_{10}\ (B)] + Rx\ NF \\ &= -174 + [10 \times \text{LOG}_{10}\ (8 \times 10^3)] + 7.59 \\ &= -127.4\ \text{dBm} \end{aligned} \tag{9.14b}$$

Now calculate the site noise:

$$N_S\ (\text{dBm}) = -174 + [10 \times \text{LOG}_{10}\ (B)] + S \qquad (9.15a)$$

where

N_S = site noise referred to the base of the receive antenna (dBm);
B = filter pass bandwidth of the susceptible receiver (Hz);
S = site noise (dB above thermal noise).

So, for this example, the site noise is:

$$\begin{aligned} N_S\ (\text{dBm}) &= -174 + [10 \times \text{LOG}_{10}\ (B)] + S \\ &= -174 + [10 \times \text{LOG}_{10}\ (8 \times 10^3)]] + 10 \qquad (9.15b) \\ &= -125\ \text{dBm} \end{aligned}$$

Now that we know both the thermal noise (N_O) and the site noise (N_S), the total noise (N_T) in the receiver's bandwidth is found by simply adding the two together.

$$N_T = 10 \times \text{LOG}_{10}\ (N_O + N_S) \qquad (9.16a)$$

where

N_T = total noise referred to the base of the receive antenna (dBm);
N_S = site noise referred to the base of the receive antenna (mW);
N_O = thermal noise power referred to the base of the receive antenna (mW).

Note again the units, site noise (N_S) and thermal noise (N_O), are in milliwatts and the total noise floor (N_T) is in dBm. This is because site noise (N_S) and thermal noise (N_O) must be converted back to a linear quantity (milliwatts) first before they can be summed together.

$$\begin{aligned} N_O\ (\text{mW}) &= 10^{(N_O\ (\text{dBm})/10)} \qquad (9.16b) \\ &= 10^{(-127.4/10)} \\ &= 1.83 \times 10^{-13}\ \text{mW} \end{aligned}$$

and

$$\begin{aligned} N_S\ (\text{mW}) &= 10^{(N_S\ (\text{dBm})/10)} \\ &= 10^{(-125/10)} \qquad (9.16c) \\ &= 3.16 \times 10^{-13}\ \text{mW} \end{aligned}$$

So

$$\begin{aligned} N_T &= 10 \times \text{LOG}_{10}\ (N_O + N_S) \\ &= 10 \times \text{LOG}_{10}\ [(1.83 \times 10^{-13}) + (3.16 \times 10^{-13})] \\ &= -123 \text{ dBm} \end{aligned} \tag{9.16d}$$

With this total noise floor, the dynamic sensitivity our receiving system is now capable of achieving in a P25 Phase 1 system using C4FM modulation designed for a minimum DAQ of 3.0 would be given by:

$$\text{Receiver dynamic sensitivity (dBm)} = N_T + (C_F\ /N + I) \tag{9.17a}$$

where

N_T = total noise referred to the base of the receive antenna (dBm);

$C_F\ /(N + I)$ = the Rayleigh faded carrier-to-noise and interference ratio (dB).

Doing the calculation:

$$\begin{aligned} \text{Receiver dynamic sensitivity (dBm)} &= -123 + (C_F\ /N + I) \\ &= -123 + 15.2 \\ &= -107.8 \text{ dBm} \end{aligned} \tag{9.17b}$$

Now let us repeat the calculations with the TTA included in the receiving system lineup. The site noise N_S will be the same, the only thing that changes is the NF used to calculate the thermal noise, N_O:

$$N_O \text{ (dBm)} = -174 + [10 \times \text{LOG}_{10}\ (B)] + Rx\ NF \tag{9.18a}$$

where

N_O = thermal noise power referred to the base of the receive antenna (dBm);

B = filter pass bandwidth of the susceptible receiver (Hz);

$Rx\ NF$ = NF of the receiving system (including TTA, antenna cabling, filters, multicoupler, and base station receiver) (dB).

So, for this example, the thermal noise is:

$$\begin{aligned} N_O \text{ (dBm)} &= -174 + [10 \times \text{LOG}_{10}\ (B)] + Rx\ NF \\ &= -174 + [10 \times \text{LOG}_{10}\ (8 \times 10^3)] + 3.11 \\ &= -131.9 \text{ dBm} \end{aligned} \tag{9.18b}$$

Adding this to the site noise to find the total noise, N_T:

$$N_T = 10 \times \mathrm{LOG}_{10}\,(N_O + N_S) \tag{9.19a}$$

where

N_T = total noise referred to the base of the receive antenna (dBm);

N_S = site noise referred to the base of the receive antenna (mW);

N_O = thermal noise power referred to the base of the receive antenna (mW).

$$\begin{aligned} N_O\ (\mathrm{mW}) &= 10^{(N_O\ (\mathrm{dBm})/10)} \\ &= 10^{(-131.9/10)} \\ &= 6.46 \times 10^{-14}\ \mathrm{mW} \end{aligned} \tag{9.19b}$$

and

$$N_S\ (\mathrm{mW}) = 3.16 \times 10^{-13}\ \mathrm{mW}\ (\text{as before}) \tag{9.19c}$$

So

$$\begin{aligned} N_T &= 10 \times \mathrm{LOG}_{10}\,(N_O + N_S) \\ &= 10 \times \mathrm{LOG}_{10}\,[(6.46 \times 10^{-14}) + (3.16 \times 10^{-13})] \\ &= -124.2\ \mathrm{dBm} \end{aligned} \tag{9.19d}$$

Now with this total noise floor, the dynamic sensitivity our receiving system in a P25 Phase 1 system using C4FM modulation designed for a minimum DAQ of 3.0, with both the TTA and the effect of site noise, would be:

$$\text{Receiver dynamic sensitivity (dBm)} = N_T + (C_F\,/N + I) \tag{9.20a}$$

where

N_T = total noise referred to the base of the receive antenna (dBm);

$C_F\,/(N + I)$ = the Rayleigh faded carrier-to-noise and interference ratio (dB).

Doing the calculation:

$$\begin{aligned} \text{Receiver dynamic sensitivity (dBm)} &= -124.2 + (C_F\,/N + I) \\ &= -124.2 + 15.2 \\ &= -109\ \mathrm{dBm} \end{aligned} \tag{9.20b}$$

As the calculations were rather long, let us get back to the point by summarizing the results in Table 9.4.

Without the effect of site noise, the TTA gave a 4.5-dB improvement in the dynamic sensitivity of the receiving system. Once site noise of 10 dB above thermal was added, the improvement dropped to just over 1 dB, which would hardly give any noticeable increase in uplink coverage at all. To sum all this up then, we can say:

- NF improvement due to the TTA does not always translate into actual sensitivity improvement; it depends on the site noise floor.
- TTAs are only useful therefore in bands where site noise is generally low. This means that TTAs are commonly used at 800 MHz, only sometimes used at UHF, and not used at VHF.

9.4 Correctly Setting the Gain of Receiver Multicoupling Systems

It is a common misunderstanding that you set up the gain of a receiver multicoupling system to cancel out the coaxial cable, filter, and splitting losses elsewhere in the system. This is not so; there is no one right way to set the gain of a multicoupling system; it depends on many factors such as the high-level interfering signals the receive antenna is exposed to and the ambient noise floor at the site. Next, we will illustrate how to best set up the gain of the multicoupling system for several common situations.

First, to show why you do not set the gain simply to overcome other system losses, let us do exactly that and see what happens to the sensitivity of the system. Let us start with the base station receiver alone, with an NF of 8 dB and a filter bandwidth of 8 kHz. Assuming a P25 Phase 1 system using

Table 9.4
Receiving System Sensitivity Showing Benefit of TTA with and without the Effect of Site Noise

Receiving System	Receiving System Sensitivity without the Effect of Site Noise	Receiving System Sensitivity with the Effect of 10-dB KTB Site Noise
Without TTA	–112.2 dBm	–107.8 dBm
With TTA	–116.7 dBm	–109 dBm

C4FM modulation designed for a minimum DAQ of 3.0, then from Chapter 6, Table 6.1, the required $C_F/(N+I)$ is 15.2 dB. In this case the dynamic sensitivity that the base station (BS) receiver can achieve on its own without the addition of any site noise would be given by:

$$\begin{aligned} &\text{BS receiver dynamic sensitivity (dBm)} \\ &= -174 + [10 \times \text{LOG}_{10}\,(B)] + Rx\,NF + (C_F/N+I) \end{aligned} \tag{9.21a}$$

where

B = filter pass bandwidth of the receiver (Hz);

$Rx\,NF$ = NF of the receiving system (including antenna cabling, filters, multicoupler, and base station receiver) (dB);

$C_F/(N+I)$ = the Rayleigh faded carrier-to-noise and interference ratio (dB).

Doing the calculation:

$$\begin{aligned} &\text{BS receiver dynamic sensitivity (dBm)} \\ &= -174 + [10 \times \text{LOG}_{10}\,(B)] + Rx\,NF + (C_F/N+I) \\ &= -174 + [10 \times \text{LOG}_{10}\,(8 \times 10^3)] + 8 + 15.2 \\ &= -111.8 \text{ dBm} \end{aligned} \tag{9.21b}$$

Let us now assume that there are four such base stations, connected to a common antenna by the simple multicoupling system shown previously in Figure 9.12. In that system, the loss of the receive antenna cable was 2 dB, the loss of the receive filter was 1.5 dB, and the loss of the 4-way splitter was 9.6 dB. This makes a total of 13.1 dB, against which the LNA gain is 27 dB, so from Figure 9.14, you can see that to cancel these out and make the overall multicoupler gain 0 dB, we need to set the variable attenuator to 14 dB.

Now to calculate the cascaded NF of this lineup:

$$\begin{aligned} &\text{Cascade NF, } F = F_1 + [(F_2 - 1)/G_1] + [(F_3 - 1)/(G_1 \times G_2)] \\ &+ [(F_4 - 1)/(G_1 \times G_2 \times G_3)] + [(F_5 - 1)/(G_1 \times G_2 \times G_3 \times G_4)] \\ &+ [(F_6 - 1)/(G_1 \times G_2 \times G_3 \times G_4 \times G_5)] \\ &= 1.585 + [(1.412 - 1)/0.631] + [(1.412 - 1)/(0.631 \times 0.708)] \\ &+ [(25.119 - 1)/(0.631 \times 0.708 \times 501)] \\ &+ [(9.12 - 1)/(0.631 \times 0.708 \times 501 \times 0.04)] \\ &+ [(6.31 - 1)/(0.631 \times 0.708 \times 501 \times 0.04 \times 0.109)] \\ &= 1.585 + 0.653 + 0.922 + 0.108 + 0.907 + 5.44 \\ &= 9.615 \end{aligned} \tag{9.22}$$

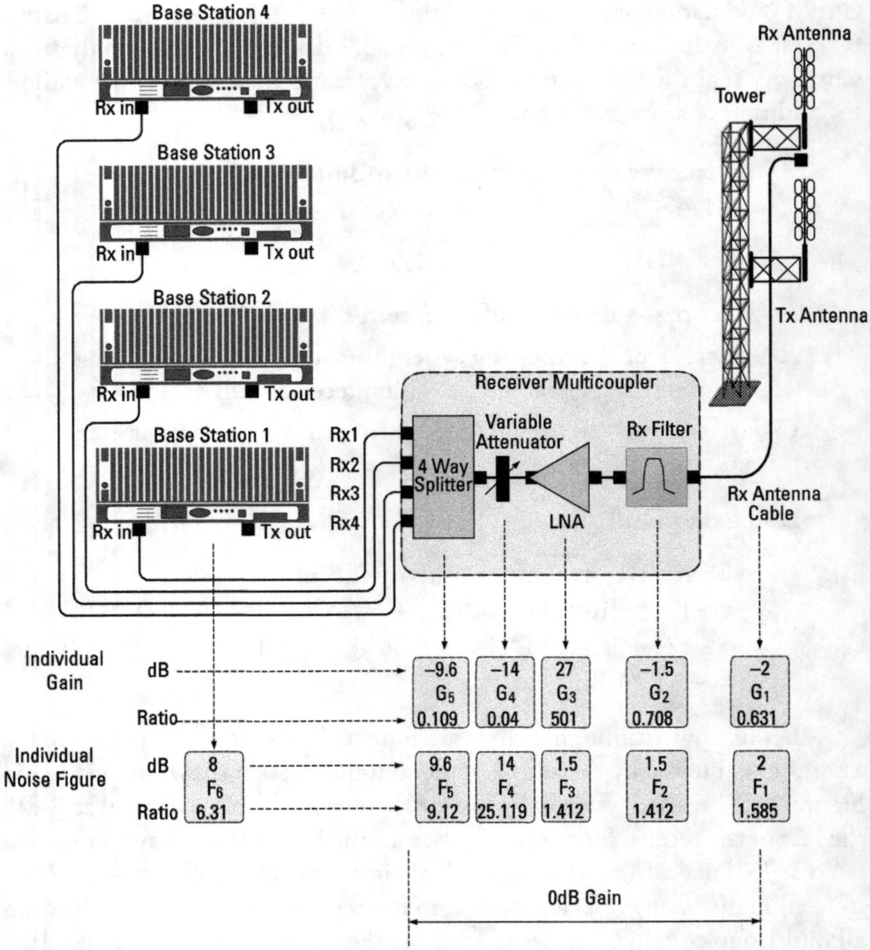

Figure 9.14 Cascaded NF of a receiver multicoupler setup for 0-dB gain.

Then convert the cascaded NF to decibels using the formula:

$$NF\,(\text{dB}) = 10 \times \text{LOG}_{10}\,(F) \tag{9.23a}$$

where F = cascade NF as a ratio.

So

$$\begin{aligned} NF\,(\text{dB}) &= 10 \times \text{LOG}_{10}\,(F) \\ &= 10 \times \text{LOG}_{10}\,(9.615) \\ &= 9.83 \text{ dB} \end{aligned} \tag{9.23b}$$

Now, the dynamic sensitivity that this receiving system can achieve without the addition of any site noise would be given by:

$$\text{Receiver dynamic sensitivity (dBm)} = -174 + [10 \times \text{LOG}_{10}(B)] + Rx\ NF + (C_F / N + I) \tag{9.24a}$$

where

B = filter pass bandwidth of the receiver (Hz);

$Rx\ NF$ = NF of the receiving system (including antenna cabling, filters, multicoupler, and base station receiver) (dB);

$C_F / (N + I)$ = the Rayleigh faded carrier-to-noise and interference ratio (dB).

Doing the calculation:

$$\begin{aligned} &\text{Receiver dynamic sensitivity (dBm)} \\ &= -174 + [10 \times \text{LOG}_{10}(B)] + Rx\ NF + (C_F / N + I) \\ &= -174 + [10 \times \text{LOG}_{10}(8 \times 10^3)] + 9.83 + 15.2 \\ &= -109.9 \text{ dBm} \end{aligned} \tag{9.24b}$$

What this means then is setting up the multicoupler gain simply to cancel out the system losses has resulted in a sensitivity that is 2 dB worse than that which was achieved by the BS receiver alone. This is because the NF of the receiving system with the attenuator set to 14 dB is 2 dB worse than that of the BS receiver alone.

Instead, let us now set the variable attenuator to 11 dB, as shown in Figure 9.15, and see what happens to the overall receiving system NF.

The gain of the multicoupler is now 3 dB, and the cascaded NF is:

$$\begin{aligned} &\text{Cascade NF, } F = F_1 + [(F_2 - 1)/G_1] + [(F_3 - 1)/(G_1 \times G_2)] \\ &+ [(F_4 - 1)/(G_1 \times G_2 \times G_3)] + [(F_5 - 1)/(G_1 \times G_2 \times G_3 \times G_4)] \\ &+ [(F_6 - 1)/(G_1 \times G_2 \times G_3 \times G_4 \times G_5)] \\ &= 1.585 + [(1.412 - 1)/0.631] + [(1.412 - 1)/(0.631 \times 0.708)] \\ &+ [(12.589 - 1)/(0.631 \times 0.708 \times 501)] \\ &+ [(9.12 - 1)/(0.631 \times 0.708 \times 501 \times 0.079)] \\ &+ [(6.31 - 1)/(0.631 \times 0.708 \times 501 \times 0.079 \times 0.109)] \\ &= 1.585 + 0.653 + 0.922 + 0.052 + 0.459 + 2.756 \\ &= 6.427 \end{aligned} \tag{9.25}$$

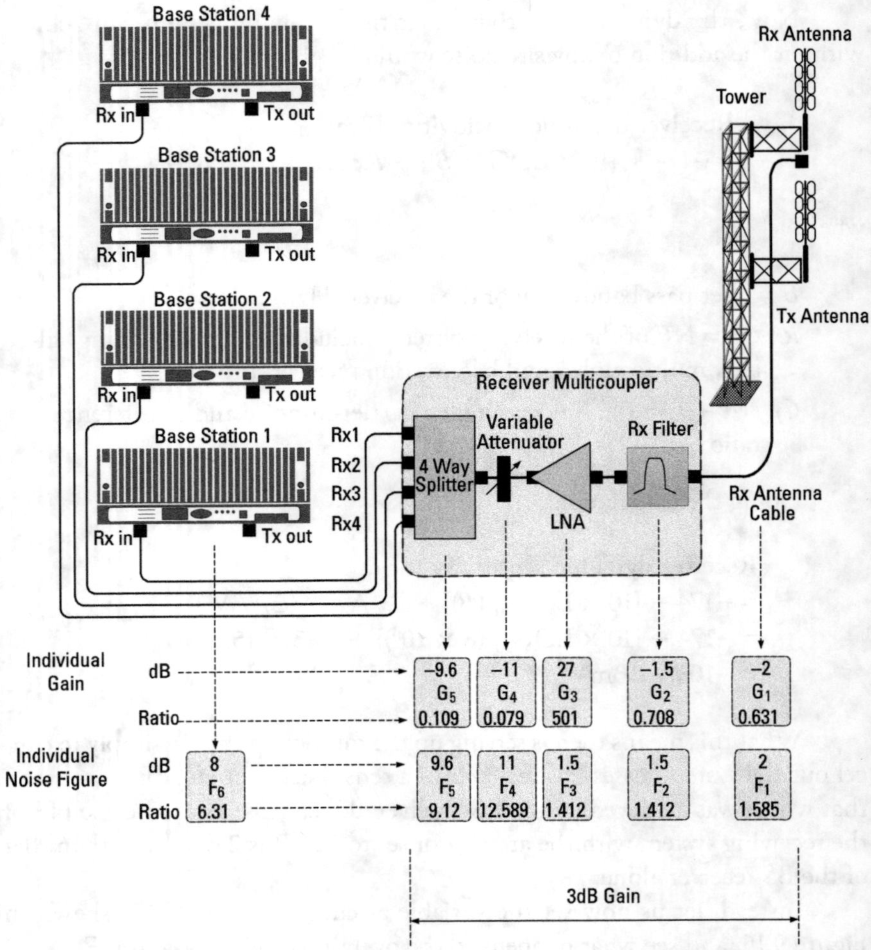

Figure 9.15 Receiver multicoupler setup to maintain the overall system NF.

Now converting this cascaded NF to decibels.

$$NF\,(\text{dB}) = 10 \times \text{LOG}_{10}\,(F) \tag{9.26a}$$

where F = cascade NF as a ratio.

So

$$\begin{aligned} NF\,(\text{dB}) &= 10 \times \text{LOG}_{10}\,(F) \\ &= 10 \times \text{LOG}_{10}\,(6.427) \\ &= 8.08\ \text{dB} \end{aligned} \tag{9.26b}$$

Now the dynamic sensitivity this receiving system can achieve without the addition of any site noise would be given by:

$$\begin{aligned}&\text{Receiver dynamic sensitivity (dBm)}\\&= -174 + [10 \times \text{LOG}_{10}\ (B)] + Rx\ NF + (C_F\ /N + I)\end{aligned} \tag{9.26c}$$

where

B = filter pass bandwidth of the receiver (Hz);

$Rx\ NF$ = NF of the receiving system (including antenna cabling, filters, multicoupler, and base station receiver) (dB);

$C_F\ /(N + I)$ = the Rayleigh faded carrier-to-noise and interference ratio (dB).

Doing the calculation:

$$\begin{aligned}&\text{Receiver dynamic sensitivity (dBm)}\\&= -174 + [10 \times \text{LOG}_{10}\ (B)] + Rx\ NF + (C_F\ /N + I)\\&= -174 + [10 \times \text{LOG}_{10}\ (8 \times 10^3)] + 8.08 + 15.2\\&= -111.7\text{ dBm}\end{aligned} \tag{9.26d}$$

This is basically the same as the −111.8-dBm sensitivity value that we calculated before for the BS receiver alone.

The point here is then:

- We do not set up the multicoupler gain to cancel out other receiving system losses.
- Rather, we set it up to maintain the NF of the overall receiving system and thereby allow one receive antenna to feed multiple receivers with no loss of sensitivity:
  - In other words, the sensitivity of each connected BS receiver is roughly the same as it would be for one BS receiver connected directly to the receive antenna.
  - This will typically mean the gain of the multicoupler system is around 3 to 4 dB.

This should always be the starting point; however (as always, especially at RF), there are situations where it pays to do things a little differently. One such case worth covering is where the site has a high ambient noise floor. To illustrate this, let us take our multicoupler from Figure 9.15 that was set up to maintain overall system NF and see how that behaves on a site where the noise floor is 10 dB above thermal as shown in Figure 9.16.

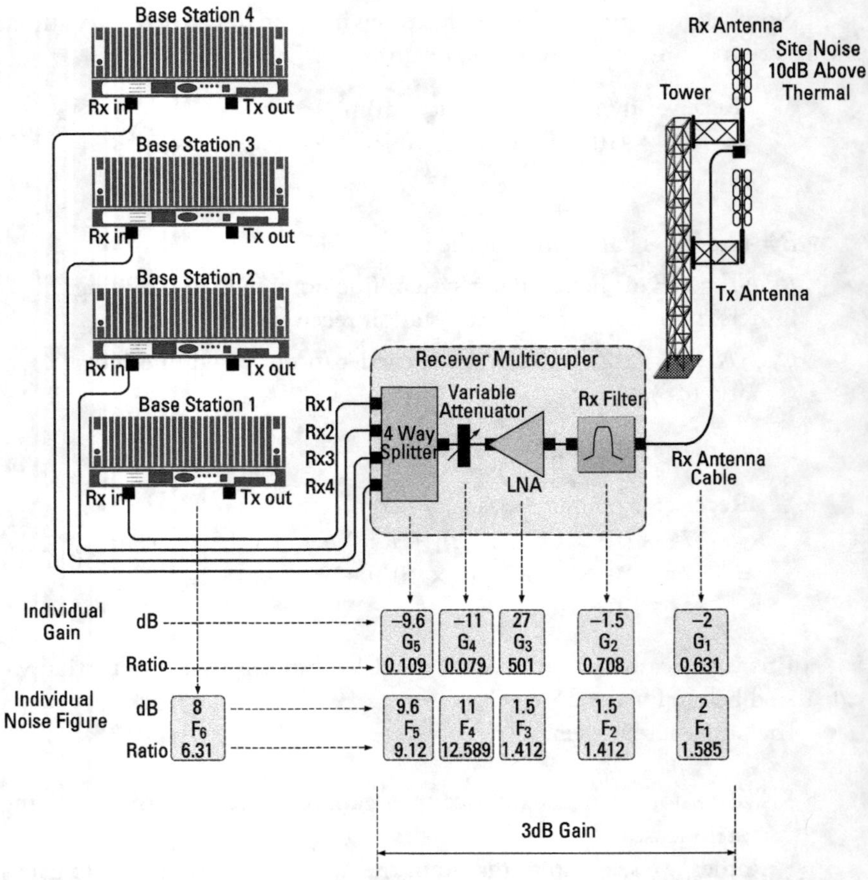

Figure 9.16 Receiver multicoupler on a site with high ambient noise floor.

To calculate the sensitivity of this multicoupler on a site with an ambient noise floor 10 dB above thermal, we start by working out the total noise, that is,

$$N_O \text{ (dBm)} = -174 + [10 \times \text{LOG}_{10} (B)] + Rx\ NF \quad (9.27a)$$

where

N_O = thermal noise power referred to the base of the receive antenna (dBm);

B = filter pass bandwidth of the susceptible receiver (Hz);

$Rx\ NF$ = NF of the receiving system (including antenna cabling, filters, multicoupler, and base station receiver) (dB).

So, for this example, the thermal noise is:

$$\begin{aligned} N_O \text{ (dBm)} &= -174 + [10 \times \text{LOG}_{10} (B)] + Rx\ NF \\ &= -174 + [10 \times \text{LOG}_{10} (8 \times 10^3)]] + 8.08 \\ &= -126.9 \text{ dBm} \end{aligned} \quad (9.27b)$$

Now calculate the site noise:

$$N_S \text{ (dBm)} = -174 + [10 \times \text{LOG}_{10} (B)] + S \quad (9.28a)$$

where

N_S = site noise referred to the base of the receive antenna (dBm);
B = filter pass bandwidth of the susceptible receiver (Hz);
S = site noise (dB above thermal noise).

So, for this example, the site noise is:

$$\begin{aligned} N_S \text{ (dBm)} &= -174 + [10 \times \text{LOG}_{10} (B)] + S \\ &= -174 + [10 \times \text{LOG}_{10} (8 \times 10^3)] + 10 \\ &= -125 \text{ dBm} \end{aligned} \quad (9.28b)$$

Adding this to the site noise to find the total noise, N_T:

$$N_T = 10 \times \text{LOG}_{10} (N_O + N_S) \quad (9.29a)$$

where

N_T = total noise referred to the base of the receive antenna (dBm);
N_S = site noise referred to the base of the receive antenna (mW);
N_O = thermal noise power referred to the base of the receive antenna (mW).

$$\begin{aligned} N_O \text{ (mW)} &= 10^{(N_O \text{ (dBm)}/10)} \\ &= 10^{(-126.9/10)} \\ &= 2.04 \times 10^{-13} \text{ mW} \end{aligned} \quad (9.29b)$$

and

$$\begin{aligned} N_S \text{ (mW)} &= 10^{(N_S \text{ (dBm)}/10)} \\ &= 10^{(-125/10)} \\ &= 3.16 \times 10^{-13} \text{ mW} \end{aligned} \quad (9.29c)$$

So

$$\begin{aligned} N_T &= 10 \times \mathrm{LOG}_{10}\,(N_O + N_S) \\ &= 10 \times \mathrm{LOG}_{10}\,[(2.04 \times 10^{-13}) + (3.16 \times 10^{-13})] \\ &= -122.8 \text{ dBm} \end{aligned} \tag{9.29d}$$

Now with this total NF, the dynamic sensitivity our receiving system in a P25 Phase 1 system using C4FM modulation designed for a minimum DAQ of 3.0 would be:

$$\text{Receiver dynamic sensitivity (dBm)} = N_T + (C_F / N + I) \tag{9.30a}$$

where

N_T = total noise referred to the base of the receive antenna (dBm);

$C_F / (N + I)$ = the Rayleigh faded carrier-to-noise and interference ratio (dB).

Doing the calculation:

$$\begin{aligned} \text{Receiver dynamic sensitivity (dBm)} &= -122.8 + (C_F / N + I) \\ &= -122.8 + 15.2 \\ &= -107.6 \text{ dBm} \end{aligned} \tag{9.30b}$$

Compared to the −111.7-dBm result obtained without site noise (see the calculation after Figure 9.15), 4 dB of sensitivity has been lost because of the elevated site noise floor. This really tells us that it is site noise that is the limiting factor determining the weakest signal that can be usably detected in this case (as opposed to the NF of the receiving system).

Strangely enough, there is actually an advantage that we can take from this if we consider the linearity of the receiving system. To revise from Chapter 3, in the same way the NF is the figure of merit relating to how much the SNR degrades through a receiving system, the intercept point is the figure of merit used to describe the linearity of the system. So, just as we can calculate the cascaded NF to understand how sensitive the receiving system might be, we can also calculate the cascaded intercept point of the receiving system in order to understand how linear it is. Let us take the receiver multicoupling system lineup of Figure 9.15 and show how to calculate its cascaded IP3.

The calculation from Figure 9.17 is done like this. Unlike the cascaded NF calculation where we start from the antenna end of the network, the cascaded intercept point calculation starts from the BS receiver and works towards the antenna. Let us start by saying the BS receiver has an input third-order intercept point (IIP3) of +16 dBm (derived from a third-order IM specification of 85 dB and the input signal levels used to measure it). This is where the calculation starts and the first thing that we do is move forward towards the antenna at one stage and work out the IIP3 at the input to the splitter by subtracting the splitter gain (which is obviously loss) of 9.6 dB, that is,

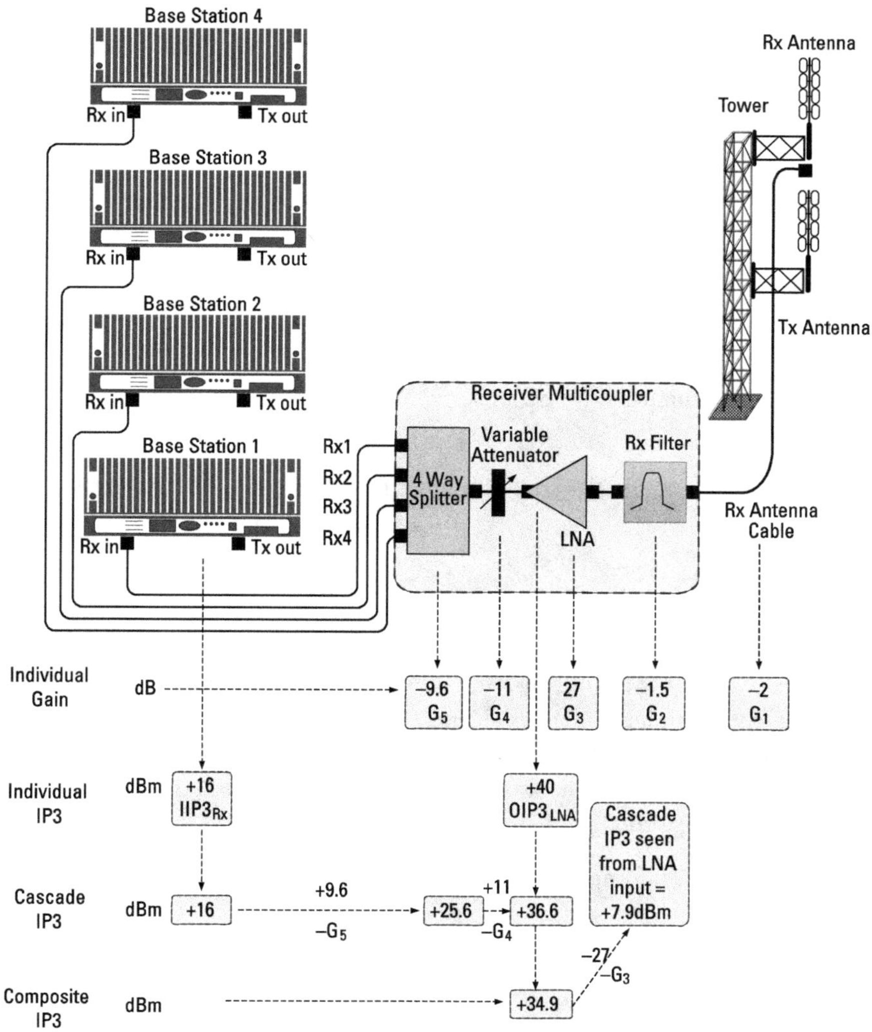

Figure 9.17 Cascaded IP3 of the receiving system from Figure 9.15.

$$\text{IIP3}_S \text{ (dBm)} = \text{IIP3}_{Rx} - G_5 \tag{9.31a}$$

where

IIP3_S = IP3 at the splitter input (dBm);
IIP3_{Rx} = IP3 at the receiver input (dBm);
G_5 = splitter gain (dB).

So

$$\begin{aligned} \text{IIP3}_S \text{ (dBm)} &= \text{IIP3}_{Rx} - G_5 \\ &= +16 - (-9.6) \\ &= +25.6 \text{ dBm} \end{aligned} \tag{9.31b}$$

Next, we move another stage towards the antenna and calculate the IIP3, referred to the input of the 11-dB attenuator.

$$\text{IIP3}_{ATT} \text{ (dBm)} = \text{IIP3}_S - G_4 \tag{9.32a}$$

where

IIP3_{ATT} = IP3 at the attenuator input (dBm);
IIP3_S = IP3 at the splitter input (dBm);
G_4 = attenuator gain (dB).

So

$$\begin{aligned} \text{IIP3}_{ATT} \text{ (dBm)} &= \text{IIP3}_S - G_4 \\ &= +25.6 + (-11) \\ &= +36.6 \text{ dBm} \end{aligned} \tag{9.32b}$$

At this point in the lineup, though, the overall IP3 is also influenced by the OIP3 of the LNA. The way this is addressed is to calculate the composite IP3 of the two. This is done by converting both the cascaded input IP3 looking into the attenuator (IIP3_{ATT}) and the OIP3 of the LNA (OIP3_{LNA}) to milliwatts and then take the reciprocal of both and add them together before taking the reciprocal of the answer; it is much like calculating the value of two resistors in parallel.

$$1/\text{Composite IP3} = 1/\text{IIP3}_{ATT} + 1/\text{OIP3}_{LNA} \tag{9.33a}$$

where

Composite IP3 = composite IP3 at attenuator input (mW);
IIP3_{ATT} = IP3 at attenuator input (mW);
OIP3_{LNA} = OIP3 of LNA (mW).

So

$$\begin{aligned} \text{IIP3}_{\text{ATT}}\ (\text{mW}) &= 10^{(\text{IIP3}_{\text{ATT}}\ (\text{dBm})/10)} \\ &= 10^{\ (+36.6/10)} \\ &= 4{,}571\ \text{mW} \end{aligned} \tag{9.33b}$$

and

$$\begin{aligned} \text{OIP3}_{\text{LNA}}\ (\text{mW}) &= 10^{(\text{OIP3}_{\text{LNA}}\ (\text{dBm})/10)} \\ &= 10^{\ (+40/10)} \\ &= 10{,}000\ \text{mW} \end{aligned} \tag{9.33c}$$

$$\begin{aligned} 1/\text{Composite IP3} &= (1/\text{IIP3}_{\text{ATT}}) + (1/\text{OIP3}_{\text{LNA}}) \\ &= (1/4{,}571) + (1/10{,}000) \\ &= 3.19 \times 10^{-4} \end{aligned} \tag{9.33d}$$

$$\begin{aligned} \text{Composite IP3 (dBm)} &= 10 \times \text{LOG}_{10}\ [1/(3.19 \times 10^{-4})] \\ &= +34.9\ \text{dBm} \end{aligned} \tag{9.34}$$

We now take this number and subtract the gain of the LNA, that is:

$$\begin{aligned} \text{IIP3}_{\text{LNA}}\ (\text{dBm}) &= \text{Composite IP3} - G_3 \\ &= +34.9 - (27) \\ &= +7.9\ \text{dBm} \end{aligned} \tag{9.35}$$

That is as far forward in the lineup as we can calculate the cascaded IP3 of the lineup as the next block; the Rx filter has a loss that depends on the input frequency, so there is no one loss figure that we can apply.

Let us now compare this with the IIP3_{LNA} of the lineup from Figure 9.14 where the attenuator was set to 14 dB. The calculation is done in exactly the same manner just covered and is represented in Figure 9.18.

The IP3 of the receiving system seen from the input to the LNA (IIP3_{LNA}) has improved to +9.8 dBm, a 2-dB improvement compared to the previous IIP3_{LNA} result of +7.9 dBm with the attenuator set to 11 dB. Note that, unlike NF, for the intercept point, a higher number is better.

Table 9.5 is probably the best way to summarize the trade-offs between sensitivity and linearity that we have shown are at play.

The point is that the sensitivity is almost the same because it is dominated by site noise rather than the system NF, but the linearity (indicated by IP3) is better with the higher attenuator setting. This makes the system more

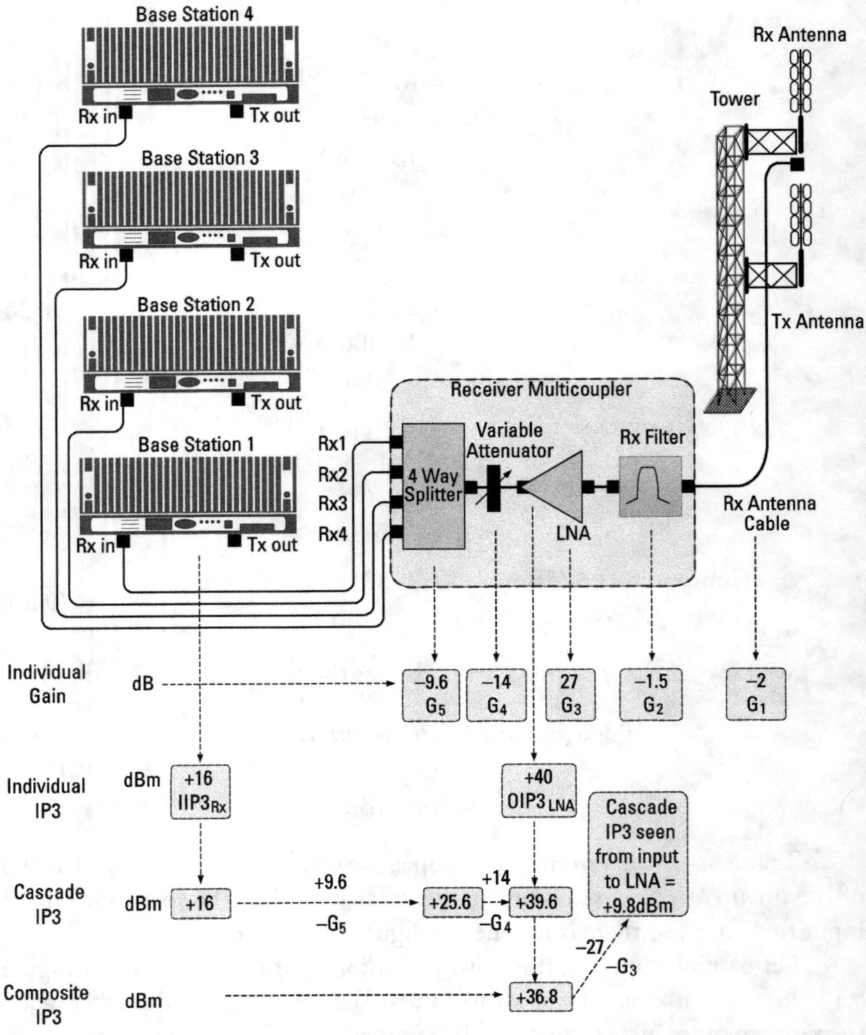

Figure 9.18 Cascaded IP3 of the receiving system with the attenuator set to 14 dB.

resilient to strong interfering signals and less likely to suffer lost coverage as a result of blocking or IM.

To summarize the important points then:

- On sites with a low ambient noise floor and little interference potential (more typical in 800-MHz LMR bands), set the gain of the

Table 9.5
Receiver Multicoupler Gain Setting

Multicoupler Attenuator Value	System NF	Sensitivity (No Site Noise)	Sensitivity (with Site Noise 10 dB above Thermal)	IP3 at LNA Input
11 dB	8.08 dB	−111.7 dBm	−107.6 dBm	+7.9 dBm
14 dB	9.83 dB	−109.9 dBm	−106.8 dBm	+9.8 dBm

multicoupler higher to maintain or improve the sensitivity of the site receiving system.

- On sites with a high ambient noise floor and significant interference potential (more typical in VHF LMR bands), set the gain of the multicoupler lower to improve the linearity of the site receiving system.

Finally, that about covers multicoupler gain setting, and you can see that there is quite a bit more to the subject than immediately meets the eye. It really does take good knowledge of the RF environment at each of the radio sites to know how to set the gain in each case and, in doing so, maximize the distance over which users can communicate using the system. This is yet one more reason why thorough, up-front RF surveys are so critical to the whole RF system design process.

9.5 Calculating Receiver Filter Requirements

We have concentrated so far on the distribution part of the receiver multicoupling system that enables us to feed several receivers from one antenna to save tower real estate. So, let us now look at other part of the site receiving system, the receiver filter, in more detail.

The receiver filter is a bandpass filter that performs two more important functions in managing interference:

- Prevents receiver blocking by rejecting high-level unwanted signals from colocated transmitters at the site.
- Prevents/manages receiver IM by providing additional rejection of specific high-level unwanted signals that would otherwise mix in the receiving system and generate on channel IM products.

Both of these functions are vitally important to ensure that none of the connected receivers are being deafened, limiting their ability to hear weak uplink signals from the field. In short, in conjunction with good frequency planning and optimal antenna system design, the receiver filter is a defense against interference from all the other transmitters at the site.

To start with then, let us calculate the rejection required in the receiver filter to prevent blocking using an example four-channel receiver multicoupler where:

- Transmitter power (per channel at the transmit antenna), $P_{\text{TX}} = +45$ dBm (or 32W). This comes from the sum of the BS transmitter output power (+50 dBm), the transmitter combiner loss (3 dB), and the RF cabling loss (2 dB).
- Transmit to receive antenna isolation is 60 dB (including the antenna gains).
- Receiver equivalent noise bandwidth, $B = 8$ kHz.
- Cascaded NF of the susceptible receiving system, $Rx\ NF = 8$ dB.
- The total site noise floor, N_T, is thermal noise only. There is no external site noise.
- Dynamic Rayleigh faded carrier-to-noise and interference ratio, $C_F / (N + I) = 15.2$ dB for a P25 Phase 1 system using C4FM modulation where a minimum DAQ of 3.0 is required (see Chapter 6, Table 6.1).
- Static carrier-to-noise ratio for the same delivered audio quality, $C_S / N = 8$ dB.

Generally, the blocking performance of a base station receiver is limited to around 105 dB (see Chapter 7). However, the 105-dB blocking performance figure is a lab measurement made with a wanted signal and an interfering signal that are both static. In practice, the wanted signal (weak uplink signal from a user in the field) will experience multipath fading, whereas the interfering signal will not, as it is coming from a colocated transmit antenna within the line of sight. To account for this, we must reduce the blocking performance by 7.2 dB, an amount equal to the difference between the static carrier-to-noise ratio (C_S / N) and the dynamic carrier-to-noise and interference ratio ($C_F / (N + I)$). The blocking performance hence becomes 97.8 dB. Also, we have four transmitters in this multichannel system, and each transposes local oscillator noise onto the receiver IF circuitry by the reciprocal mixing mechanism. We can account for this by reducing the blocking performance by a further 6 dB, to 91.8 dB. In addition, the BS receivers are also subject to sideband noise from the transmitters, so in effect there are two on-channel noise sources

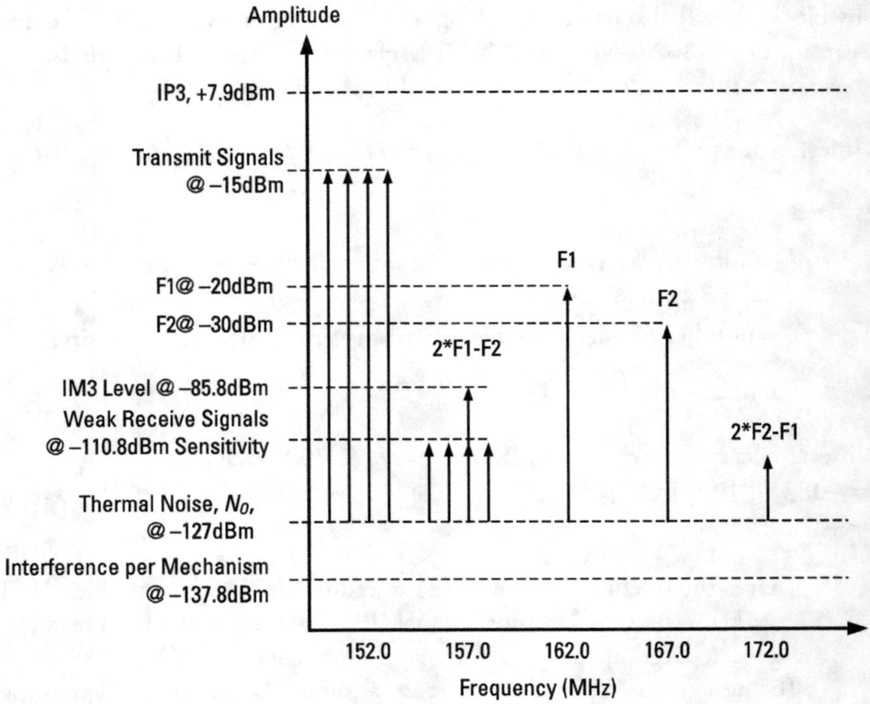

Figure 9.23 Third-order receiver IM levels.

Table 9.6
Example Receiver Filter Specifications

Parameter	**Performance**
Filter characteristic	Bandpass
Passband	155.0 (F_L) – 158 MHz (F_H)
In-band insertion loss	<1.5 dB
In-band return loss	Better than 22 dB
Blocking rejection (150.0 to 153.0 MHz)	>12 dB
IM rejection (162.0 MHz)	>20 dB
IM rejection (167.0 MHz)	>27 dB
Stopband	$<F_L - 10$ MHz, $> F_H + 10$ MHz
Minimum rejection in stopband	>70 dB

present. We can account for this by reducing the blocking performance by another 3 dB. Hence, the blocking performance becomes 88.8 dB. What that means is the maximum transmit signal level that we can tolerate at any BS receiver connected to the multicoupler is 88.8 dB above the minimum wanted receive signal.

The next step towards working out the required receiver filter rejection is to calculate the minimum wanted receive signal level (i.e., the receiver sensitivity). As we are well versed by now, the calculation starts with the receiver thermal noise floor (N_O).

$$\text{Receiver thermal noise floor, } N_O \text{ (dBm)} = -174 + [10 \times \text{LOG}_{10}(B)] + Rx\,NF \tag{9.36}$$

where

N_O = thermal noise power at the receive antenna (dBm);

B = filter pass bandwidth of the receiver (Hz);

$Rx\,NF$ = BS receiver NF (dB).

So, for our receiving system, we get a thermal noise floor of:

$$\begin{aligned} N_O \text{ (dBm)} &= -174 + [10 \times \text{LOG}_{10}(8 \times 10^3)] + 8 \\ &= -127 \text{ dBm/8 kHz} \end{aligned} \tag{9.37}$$

Remember that there is no external site noise in this example, so to find the total noise floor (N_X), we simply add the 1-dB allowable receiver desensitization to the thermal noise floor (N_O).

$$N_X = N_T + 1 \text{ dB} \tag{9.38}$$

where

N_X = total noise and interference floor referred to the base of the receive antenna (dBm);

$N_T = N_O$ = thermal noise power referred to the base of the receive antenna (mW).

So

$$\begin{aligned} N_X &= N_T + 1 \text{ dB} \\ &= -127 + 1 \\ &= -126 \text{ dBm} \end{aligned} \tag{9.39}$$

To find the receiver sensitivity, we add the relevant $C_F/(N+I)$ threshold to the total noise floor (N_O); this is how far the received signal must be above the noise and interference in a real-world Rayleigh faded environment in order to provide the minimum acceptable level of link performance. If we assume that our system is a P25 Phase 1 system using C4FM modulation and a minimum DAQ of 3.0 is required, then the receiver sensitivity becomes:

$$\begin{aligned}\text{Receiver sensitivity (dBm)} &= N_X + [C_F/(N+I)] \\ &= -126 + 15.2 \\ &= -110.8 \text{ dBm}\end{aligned} \tag{9.40}$$

where

Receiver sensitivity = the received signal level required for minimum performance (dBm);

N_X = total noise and interference floor referred to the base of the receive antenna (dBm);

$C_F/(N+I)$ = the Rayleigh faded carrier-to-noise and interference ratio (dB) (see Chapter 6, Table 6.1).

The maximum interfering signal level that we can therefore tolerate then at the base of the receive antenna is the sum of the above, the minimum wanted receive signal, and the blocking performance.

$$\begin{aligned}&\text{Maximum interfering signal at receive antenna (dBm)} \\ &= \text{Receiver sensitivity} + \text{Blocking} \\ &= -110.8 + 88.8 \\ &= -22 \text{ dBm}\end{aligned} \tag{9.41}$$

Next, we need to know the level of the interfering transmit signal (at the base of the receive antenna) that we need to reject. On shared radio sites, there may be many high-level signals from colocated transmitters present, but, for this example, let us just consider the transmit signals from our associated transmitter combiner. At the beginning of this example, we stated that the transmitter power per channel at the transmit antenna input was +45 dBm and that the transmit to receive antenna isolation was 60 dB. So, the level of each transmit signal at the receive antenna output would be −15 dBm. Also, we just calculated the maximum interfering signal level that we can tolerate at the receive antenna to be −22 dBm for 1 dB of receiver desensitization. So, the rejection required in the receive filter at this transmit frequency is simply the difference between the two.

Receive filter rejection at transmit frequency (dB)
= Interfering signal level at receive antenna (dBm)
− Maximum interfering signal level at receive antenna (dBm) (9.42)
= −15 − (−22)
= 7 dB

This relatively easy requirement really emphasizes the benefit of maximizing the transmit to receive antenna isolation, which is effectively providing most of the blocking rejection for free.

What is going on is much easier to show in a diagram like Figure 9.19.

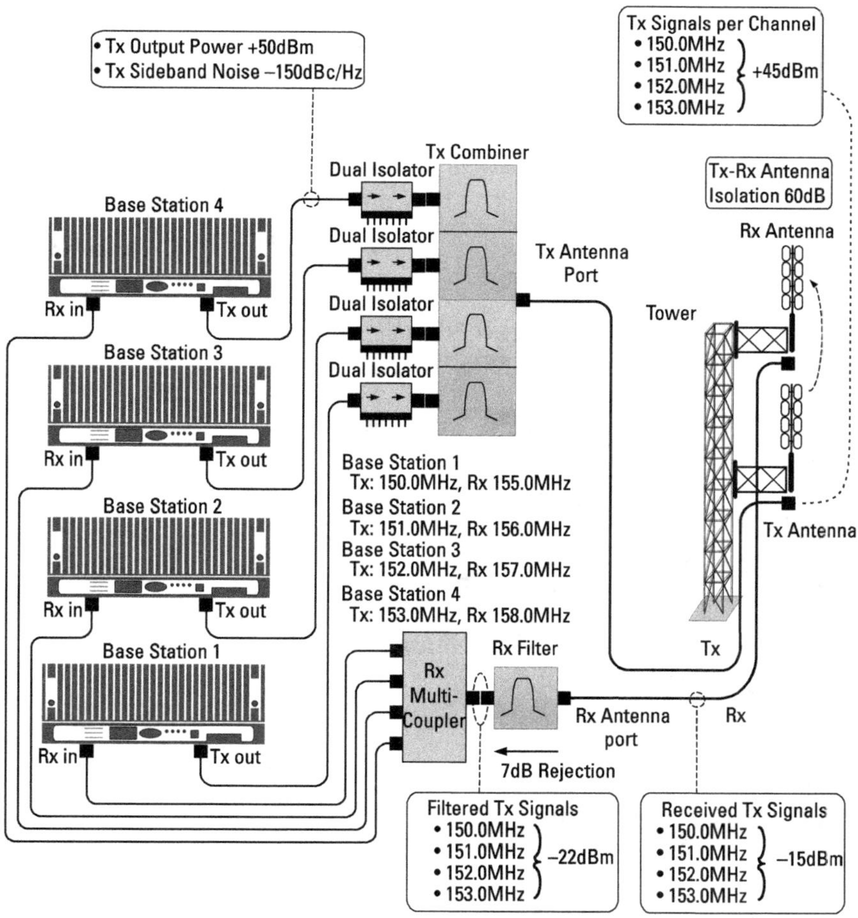

Figure 9.19 Receiver filter blocking rejection.

As stated, this example is just for associated transmitter signals. The blocking rejection required in the receiver filter for unrelated high-level signals from other colocated transmitters at the site can be calculated in the same way; all you have to know is their level at the receive antenna.

Now consider what happens if two external interfering signals are picked up from other transmitters on the site: one at 162 MHz at a level of –20 dBm, and another at 167 MHz at a level of –30 dBm, as shown in Figure 9.20.

The problem that these external signals cause is that they generate a third-order IM product at the BS 3 receive frequency of 157.0 MHz, as shown in Figure 9.21.

To see how to deal with this, let us start with revising the receiver multicoupler of Figure 9.17 that was set up for an 8-dB overall system NF, with the two external interfering signals applied as shown in Figure 9.22.

Thinking about the level of the $2 \times F_1 - F_2$ product that lands on the 157.0-MHz receive frequency, notice that the level of the external interfering

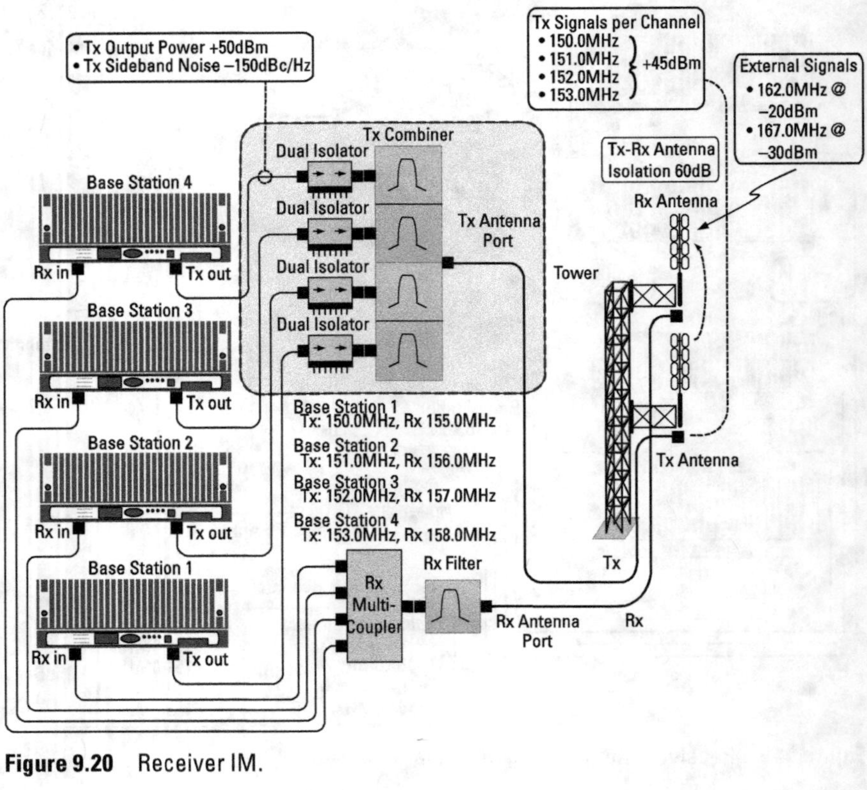

Figure 9.20 Receiver IM.

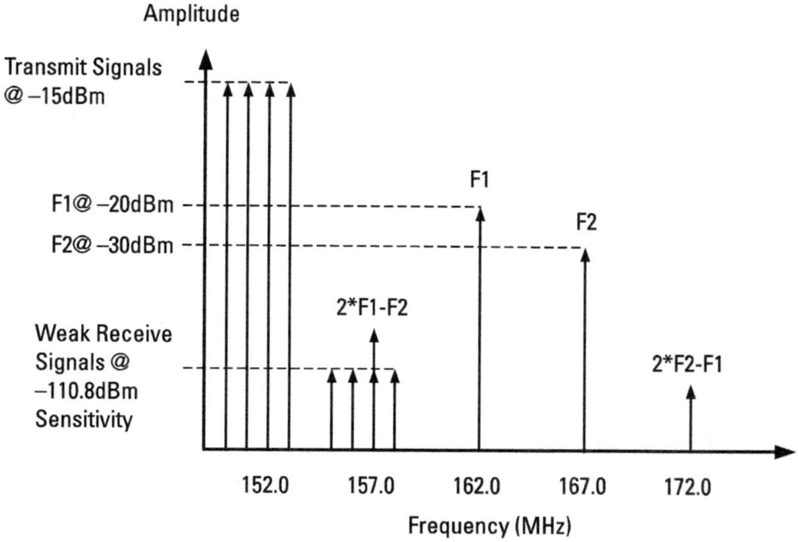

Figure 9.21 Third-order receiver IM.

signal F_1 (−20 dBm) is 27.9 dB below the IP3 at the output of the receiver filter (+7.9 dBm). So, because the order of F_1 involved in this IM product is the second harmonic, the level of the $2 \times F_1 - F_2$ product drops by twice 27.9 dB below the IP3, to −47.9 dBm. In addition, the level of external interfering signal F_2 (−30 dBm) is 37.9 dB below the IP3 at the output of the receiver filter (+7.9 dBm), so because the order of F_2 involved in this IM product is the fundamental, the level of the $2 \times F_1 - F_2$ product drops by another 37.9 dB below the IP3 at the output of the receiver filter to −85.8 dBm. This is way above our desired sensitivity of −110.8 dBm and our calculated noise floor of −126 dBm. What this means in reality is that the IM product is causing strong interference that will severely limit the distance from which a user can communicate with the site on the BS 3 receive frequency of 157.0 MHz.

The next step is to calculate the interference threshold (I_T), the level below which the sum of any transmitter harmonic, transmitter IM, transmitter sideband noise, receiver blocking, and receiver IM contributions must be suppressed at the base of the receive antenna so as not to result in any more than 1 dB of desensitization of any of the connected receivers. How to calculate the interference threshold was covered fully in Chapter 6, but we can easily do it empirically here, like this. Earlier in this example, we calculated that the receiver thermal noise floor (N_O) was −127 dBm. On top of this, we allow the total interference to cause a further 1-dB rise. So a useful rule of

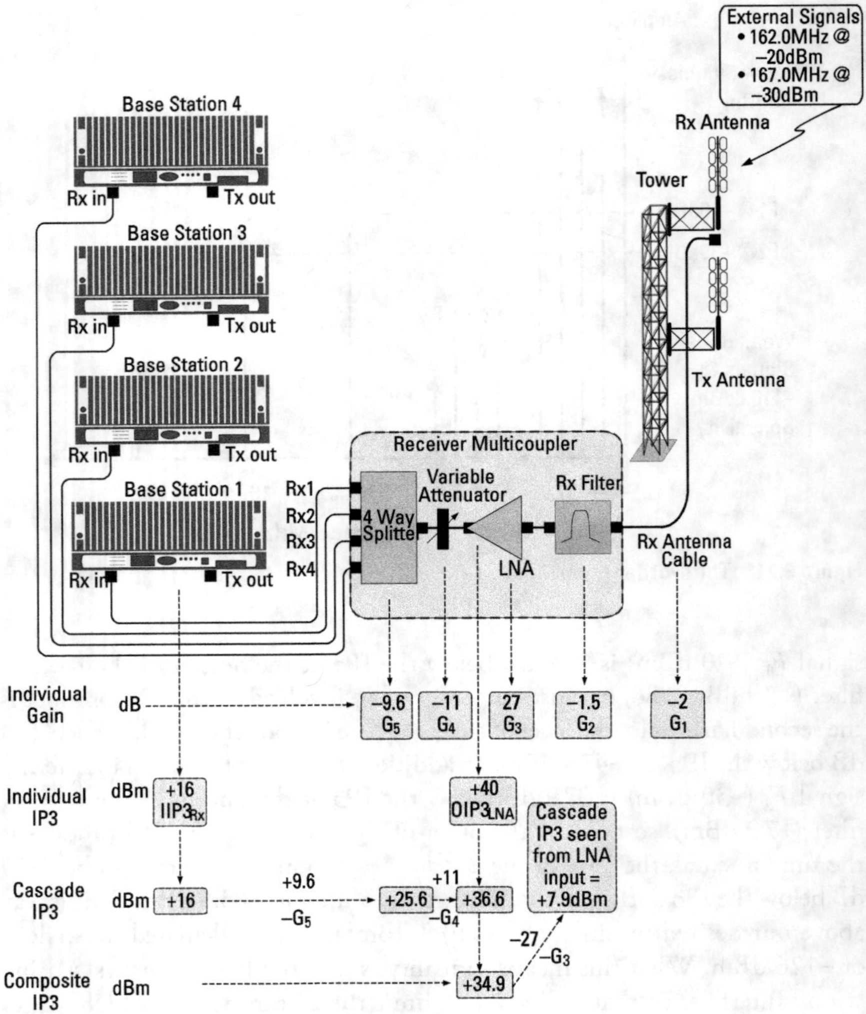

Figure 9.22 Receiver multicoupler for an IM example.

thumb to use here is if you take a noise source in dBm and add another noise source that is 6 dB lower, then the total noise increases by 1 dB. Hence, if we take our receiver noise floor of −127 dBm, then a further 1-dB degradation is caused by adding another noise source (i.e., at the interference threshold, I_T), which is 6 dB lower, or −133 dBm. That is the total interference threshold, but we also need to remember that we have multiple interference mechanisms contributing to this 1-dB rise. Transmitter sideband noise and noise transposed by blocking will also be present along with the receiver IM identified here as a result of the 162.0-MHz and 167.0-MHz external interfering signals. We have three interference mechanisms at play; therefore:

$$\text{Interference per mechanism (dBm)}, I'_T = I_T - [10 \times \text{LOG}_{10}(n)] \quad (9.43\text{a})$$

where

I_T = interference threshold, the total interference level at the base of the Rx antenna (dBm);

n = number of interference mechanisms simultaneously occurring.

So, in this case:

$$\begin{aligned} \text{Interference per mechanism (dBm)}, I'_T &= I_T - [10 \times \text{LOG}_{10}(n)] \\ &= -133 - [10 \times \text{LOG}_{10}(3)] \quad (9.43\text{b}) \\ &= -137.8 \text{ dBm} \end{aligned}$$

That is the level to which we need to reduce the IM product identified at 157.0 MHz to prevent any more than 1-dB deafening of the BS 3 receiver.

Again, a diagram such as Figure 9.23 represents this best.

The task at hand then is to define the additional rejection needed in the receiver filter to suppress the $2 \times F_1 - F_2$ IM product down to the interference per mechanism threshold of −137.8 dBm.

Let us start by considering the rejection of F_1 as that is the interfering signal closest to the receiver filter passband of 155 to 158 MHz. If we design the receiver filter to provide 15 dB of rejection at F_1, that would suppress the level of the $2 \times F_1 - F_2$ IM product by twice that amount (because it is the second harmonic of F_1 that is involved) to −115.8 dBm. If we then design the receiver filter to also provide 22 dB of rejection at F_2, then the level of the IM product reduces further by the same amount (because the fundamental of F_2 is involved) to the interference per mechanism threshold (I_T) of −137.8 dBm. The job is done.

So, now if we combine what we have learned, we have the basis of a specification for the receiver filter to mitigate both blocking and IM, as per Table 9.6. Note that an extra 5 dB has been added to all the calculated rejection figures to account for temperature drift and aging.

This is better represented in Figure 9.24.

This example shows the case where no interfering signals exist between the receive frequencies; hence, a common receive window filter can be used. If an interfering signal was present in between the receive signals, then a

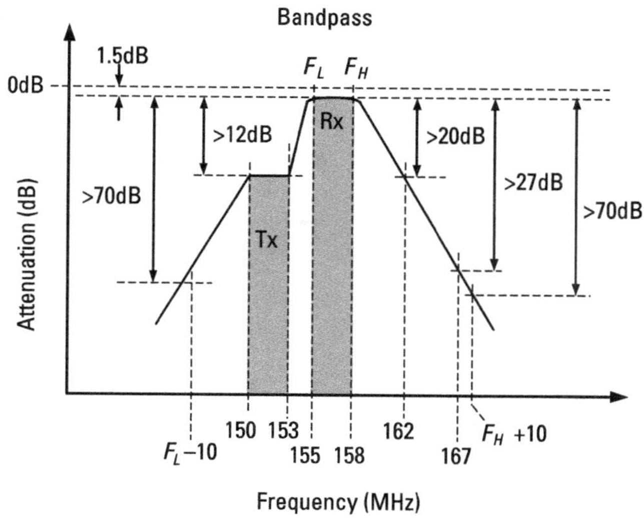

Figure 9.24 Receiver filter frequency response.

channelized receiver filter would be required and the rejection requirements would need to be calculated individually for each of them using the same method that has just been presented.

9.6 Differences for Duplexed Multichannel Systems

Despite strongly recommending the use of separate transmit and receive antennas for multichannel RF systems, sometimes a single duplexed antenna has to be used due to an existing tower structural loading or lack of suitable locations on the tower for two antennas.

However, a duplexed multichannel system still needs a receiver multicoupler, for which the design considerations and requirements are much the same as those described throughout this chapter. There are a couple of key differences, though. First, the lack of any antenna isolation means that there is an increased risk of passive IM and a subsequent loss of uplink coverage. This is especially true if the IM mechanisms involve the system's own transmitter frequencies, which are present at full power on the duplexed antenna. In this case, IM is generated at much higher levels, meaning that high-order products can be present on the antenna at levels that will deafen a receiver sharing the duplexed antenna. As a result, with duplexed multichannel systems, it is necessary to carry out IM analysis to well beyond the seventh order. My own practical experience dictates that, for duplexed multichannel systems, receiver IM searches need to be carried out to the twenty-first order, which makes frequency planning much harder.

Second, all the suppression necessary to protect the connected receivers against blocking and IM now has to be achieved in the receiver filter. For example, in the receiver blocking calculation presented earlier in this chapter (see Figure 9.19), the blocking rejection that we would need from the receive filter would be 67 dB if it was a duplexed system, as opposed to the 7 dB that we calculated with a 60-dB isolation between separate transmit and receive antennas.

9.7 Site Receiver Multicoupler Design Rules

Before concluding with some best-practice receiver multicoupler design rules, it is worth reminding ourselves of the points made in this chapter; that the proper design of the receiver multicoupler is vital and gives the benefits of maximum coverage and minimum interference potential.

The following are general best-practice design guidelines that describe how to design optimal receiver multicouplers for deployment on radio towers.

1. Use separate transmit and receive antennas for multichannel systems. Do not use duplexers for multichannel RF system designs. Sometimes it is unavoidable due to tower congestion or loading, but, as a general rule, avoid multichannel duplexed RF systems.
2. Use a standard receiver multicoupler when there are no high-level transmitters at the site with frequencies that fall in between the highest and lowest frequencies of the receivers connected to the multicoupler.
3. A channelized receiver multicoupler must be used when there are high-level transmitters at the site that fall in between the highest and lowest receive frequencies of the receivers connected to the multicoupler. A combination of these two types of multicouplers can be used if some, but not all, of the connected receivers can be fed from a common receiver filter without allowing high-level transmit carriers through.
4. Filtering in receiver multicoupling systems must be predominantly bandpass.
   - Never use band-reject (notch) filtering alone in receiver multicouplers.
   - A combination of bandpass and band-reject filtering is acceptable to provide extra suppression at specific (usually very close) transmit frequencies.

5. The multicoupler must have enough gain and low enough NF to enable the required dynamic receiver sensitivity to be achieved.
   - A receiver multicoupler with one LNA can be used if the receiving system losses (e.g., splitters) are such that the required dynamic receiver sensitivity can be achieved.
   - If the required dynamic receiver sensitivity cannot be achieved with one LNA, use two LNAs with the variable attenuator between them. The first LNA should be optimized for the NF, and the second LNA should be optimized for linearity (high IP3).
6. The receiver filter shall be designed so that no more than 1 dB of desensitization of any site receiver occurs due to receiver blocking. Normally, we can consider that it is the transmit signal closest to the receive frequency that causes the main blocking problem. Once we have determined the attenuation required in the receiver filter at the closest transmit frequency, all other transmit frequencies are further away and so will see a greater attenuation through the receiver filter and will not noticeably contribute:
   - Where there are multiple high-level carriers at similar levels and offsets away from the receive frequency, the blocking performance of the receiver decreases by 6 dB for two transmit signals.
   - The chances of constructive correlation between these noise sources becomes less, the more sources there are. If there are 3 or more high-level transmit carriers at similarly high levels and similar offsets away from the receive frequency, 9 dB is subtracted from the receiver blocking performance.
7. The receiver filter shall be designed so that no more than 1 dB of desensitization of any site receiver occurs due to receiver IM.
   - Receiver IM analysis must be carried out including all site transmit signals present at the receive antenna at a level of –40 dBm or stronger.
   - For multichannel systems with separate transmit and receive antennas with 60 dB of isolation between them, receiver IM analysis shall be carried out to the seventh order within a search bandwidth of:

$$\text{Search bandwidth (kHz)} = (n + 1) \times \text{Channel bandwidth} \quad (9.44)$$

where n = maximum order of IM to be searched for.

For seventh-order IM and a channel bandwidth of 12.5 kHz (typically for LMR systems), the search bandwidth is 100 kHz (i.e., 50 kHz either side of each receive frequency).

- If a duplexed multichannel system with a common transmit and receive antenna is unavoidable, transmitter IM analysis shall be carried out to the twenty-first order within a search bandwidth of:

$$\text{Search bandwidth (kHz)} = (n + 1) \times \text{Channel bandwidth} \qquad (9.45)$$

where n = maximum order of IM to be searched for.

For the twenty-first-order IM and a channel bandwidth of 12.5 kHz (typically for LMR systems), the search bandwidth is 275 kHz (i.e., 137.5 kHz either side of each receive frequency).

8. The RF connector types used on receiver multicouplers shall be:
   - Receiver filter:
     - The input connector of the receiver filter shall be 7/16 DIN Female (to minimize PIM).
     - The output connector of the receiver filter shall be either 7/16DIN Female or type N Female.
   - Receiver multicoupler (LNA, attenuator, splitter unit):
     - The input connector of the receiver multicoupler shall be type N Female.
     - The output connector of the receiver filter shall be either type N Female or BNC Female.

Selected Bibliography

RFI Technology Solutions, 380-520MHz Receiver Multicoupler data sheet, https://www.rfi.com.au/core/media/media.nl?id=1930503&c=3389566&h=CE_f2kNckoJDG7gXr7LVUNfzn5X7qOvhVm43ScD2sQpf16gW.

RFI Technology Solutions, 400-520MHz UHF Preselector Filter data sheet, https://www.rfi.com.au/core/media/media.nl?id=3897414&c=3389566&h=AXCFTiyrQKADwP2myLnnM44CsOiwlyP7BfThoYJc8yE1qpOZ.

RFI Technology Solutions, 700-800MHz Tower Top Amplifier System data sheet, https://www.rfi.com.au/core/media/media.nl?id=2432526&c=3389566&h=M96_zhTAF1FGGarDZM6WkGKZyGlPKuYWwRQNEx6YSN8JkpnO.

Telecommunications Industry Association, Report TSB-88.1-F, Wireless Communications Systems—Performance in Noise and Interference Limited Situations—Part 1: Recommended Methods for Technology Independent Narrowband Performance Modelling, April 2022.

10

Power and Antenna System Monitors

Combining a number of transmitters onto a common antenna does cause one annoying problem: antenna VSWR detection.

Most LMR base station transmitters include the capability to measure forward and reverse power to detect when there is a problem with the antenna. However, the necessary use of isolators in transmitter combiners means any reverse power is now routed to the isolator dump port instead of the base station. In other words, the antenna could be damaged, and yet the base station VSWR detection circuitry is unable to detect that there is anything wrong.

10.1 Power Monitors

The first and simplest solution to this problem is to include a power monitor in the system as shown in Figure 10.1. These are low-loss units based on an internal 30-dB directional coupler (chosen to minimize signal losses) connected immediately after the transmitter combiner output. The forward-and-reverse power sensed by this coupler are then processed in the power monitor unit to provide the measurement of:

- Composite transmitter forward and reverse power (the sum of all the active transmitters combined);
- Transmit antenna VSWR.

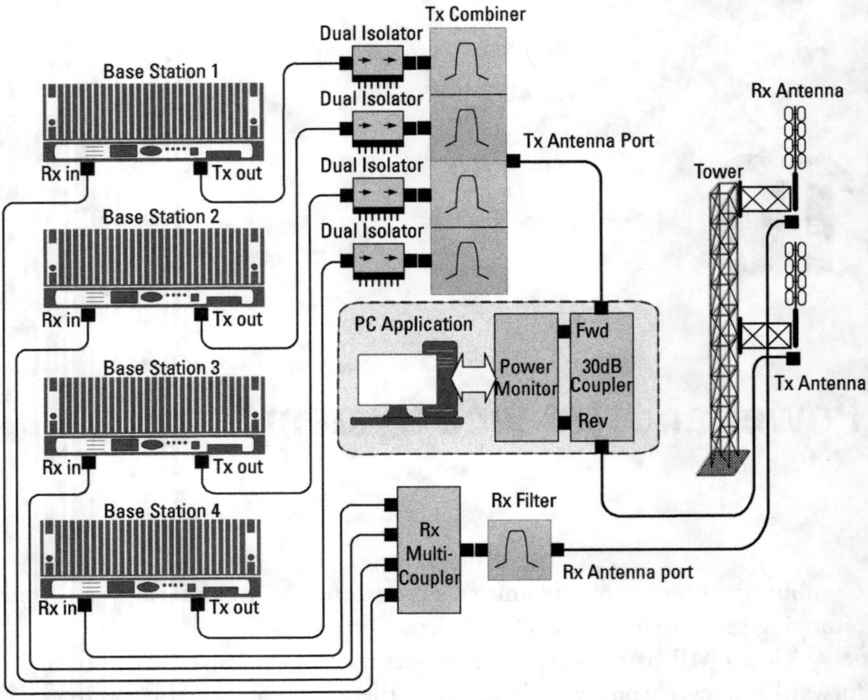

Figure 10.1 Composite power monitor.

The internet connection then allows for remote monitoring of these measurements via a simple PC application. If the reverse power or VSWR rise above a predetermined programmed level, then an alarm is activated.

10.2 Antenna System Monitors

Antenna system monitors are a more capable variation on this theme. More capability comes at a cost of greater complexity, but the idea is still relatively simple. As illustrated in Figure 10.2, the system again begins with a 30-dB directional coupler connected at the transmitter combiner output. Unlike most power monitors, the coupler is not part of the antenna system monitoring unit and must be purchased separately. The forward-and-reverse ports of the coupler are connected to the antenna system monitor for processing.

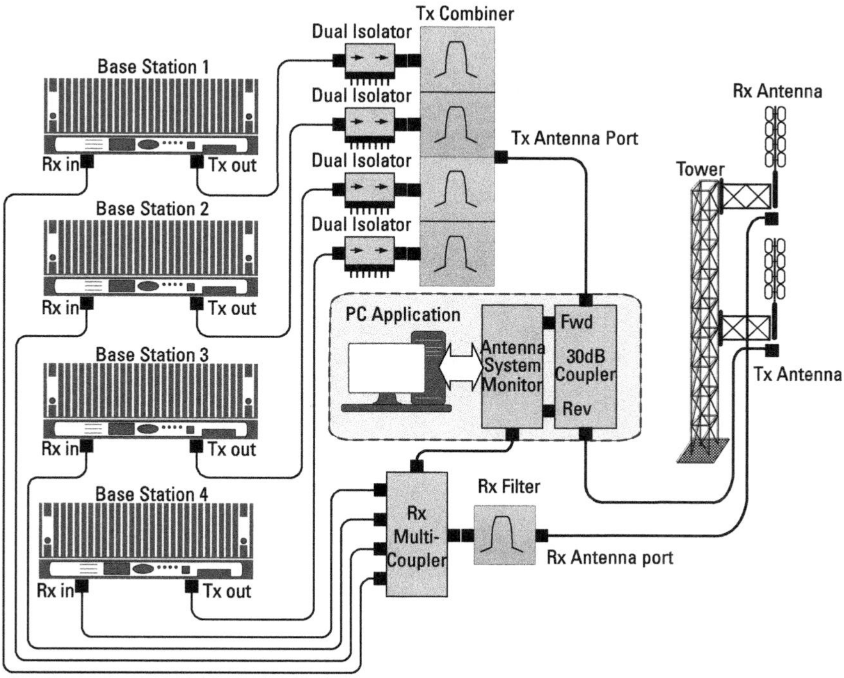

Figure 10.2 Channelized antenna system monitor.

Additionally, a dedicated receive port is connected to a spare receiver multicoupler output.

Together, this provides the capability to remotely monitor:

- Channelized transmitter forward-and-reverse power (unlike the power monitor, the power can be separately reported for each individual transmitter, as opposed to just the total combined power);
- Transmit antenna VSWR;
- Transmit to receive antenna isolation;
- Receiver signal level (RSSI).

The benefit of including power or antenna system monitors in your repeater site RF combining systems is that you get early warning when the transmit antenna starts to degrade. That enables investigation and repair to take place before the antenna fails completely and causes a sudden loss of coverage (which would equal lost operation, productivity, and money).

References

[1] dBSpectra, DSXPM5 Power Monitor data sheet, http://dbspectra.wpengine.com/specifications/DSXPM5Series_SMART_Transmit_Power_Monitor.pdf.

[2] RFI Technology Solutions, Antenna System Monitor data sheet, https://www.rfi.com.au/core/media/media.nl?id=2490452&c=3389566&h=EKeZ39Wc7o6Dcgrxosqr_8BnkfspEBVj7-I3rub-7trVB5Ib.

11

Transmission Line Systems

Now that the antenna system, transmitter combiner, and receiver multicoupler designs have been addressed, there is one final part of the design: the RF cables and connectors used to connect these subsystems together. This part of the design is often overlooked, but is worth covering in detail because the appropriate choice of RF cables and connectors has a beneficial effect on the system performance.

In this chapter, we will look at the various types of transmission lines and connectors, where to use them, and cover related items such as surge suppressors and grounding devices.

11.1 Transmission Line Systems for Multichannel Site RF System Designs

Let us start by looking at the RF cables involved in connecting up a multichannel site RF system with separate transmit-and-receive antennas. Figure 11.1 shows a typical multichannel RF system cabling configuration, and then Table 11.1 gives the specifications for each of the cables.

As a general note, Superflex cables have a smaller bend radius and are used for convenience in the equipment rack to connect the base stations to the transmitter combiner and receiver multicoupler. However, they are more

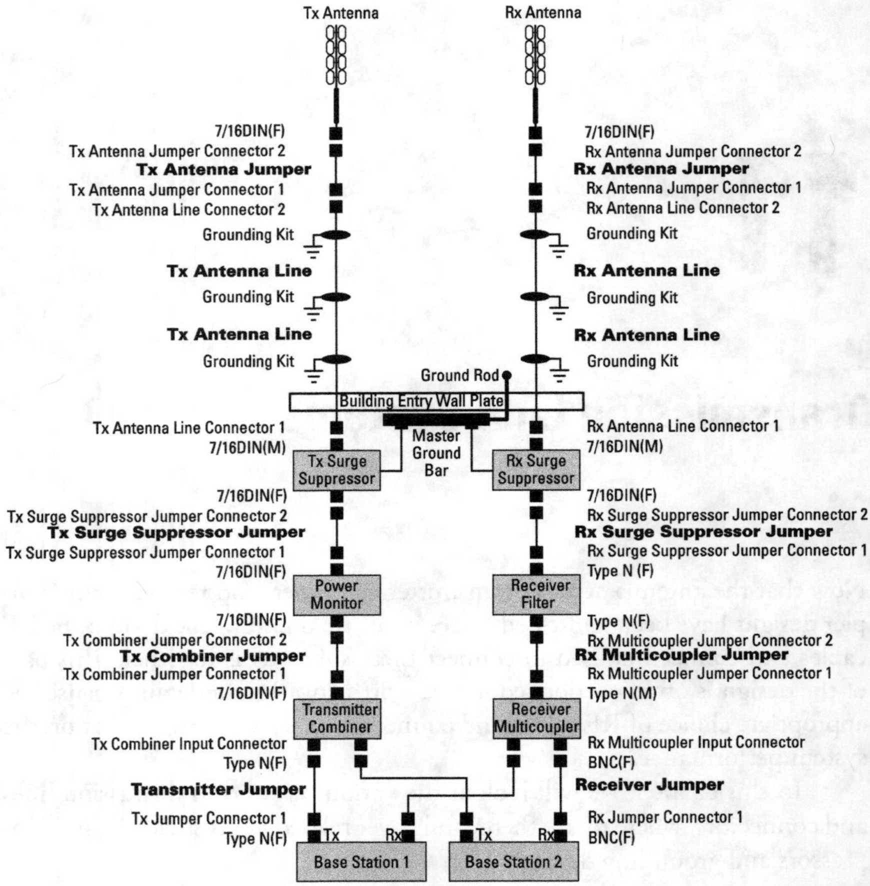

Figure 11.1 Transmission lines for multichannel site RF system.

prone to damage, and so from the output of the transmitter combiner onwards to the antenna (where multiple signals are present and many of the cables are outside), more durable Heliax cables are used to minimize the risk of damage and subsequent generation of passive IM.

11.2 Transmission Line Systems for Single-Channel Duplexed RF System Designs

Now let us move on to look at the RF cables involved in connecting up a single-channel site RF system with a duplexed antenna. Figure 11.2 shows a

Table 11.1
Transmission Line Specifications for Multichannel Site RF System

Cable Name	Cable Type	Example Cables	Connector 1	Connector 2
Tx Antenna Jumper, Rx Antenna Jumper	1/2-inch Heliax (see Note 1)	Commscope LDF4-50, RFS LCF12-50J	7/16DIN (M)	7/16DIN (M)
Tx Antenna Line, Rx Antenna Line	7/8-inch Heliax (see Note 2), 1 1/4-inch Heliax	Commscope AVA5-50, RFS LCF78-50, Commscope AVA6-50, RFS LCF114-50	7/16DIN (F)	7/16DIN (F)
Tx Surge Suppressor Jumper	1/2-inch Heliax	Commscope LDF4-50, RFS LCF12-50J	7/16DIN (M)	7/16DIN (M)
Rx Surge Suppressor Jumper	1/2-inch Heliax	Commscope LDF4-50, RFS LCF12-50J	Type N(M)	7/16DIN (M)
Tx Combiner Jumper	1/2-inch Heliax	Commscope LDF4-50, RFS LCF12-50J	7/16DIN (M)	7/16DIN (M)
Rx Multicoupler Jumper	1/4-inch Superflex	Commscope FSJ1-50, RFS SCF14-50	Type N(M)	Type N(M)
Transmitter Jumper	1/2-inch Superflex	Commscope FSJ4-50, RFS SCF12-50	Type N(M)	Type N(M)
Receiver Jumper	1/4-inch Superflex	Commscope FSJ1-50, RFS SCF14-50	BNC (M)	BNC (M)

Notes: (1) The antenna jumpers at the end of the Tx and Rx antenna lines are made from thinner ½-inch cable to relieve pressure on the antenna connectors. Both the Tx and Rx antenna jumpers are generally kept short, about 10 ft in length, to minimize their effect on system losses. (2) For the Tx and Rx antenna lines, the 7/8-inch-diameter Heliax cable is used if the total length is less than 200 ft. If the length is greater than 200 ft, then 1¼-inch Heliax is used to minimize losses.

typical single-channel RF system cabling configuration, and then Table 11.2 gives the specifications for each of the cables.

11.3 RF Surge Suppression Devices

As many radio repeater sites are located at geographic high points to maximize coverage, it means the antennas are prone to lightning strikes. If left unprotected, all the connected RF equipment would be destroyed in such an event

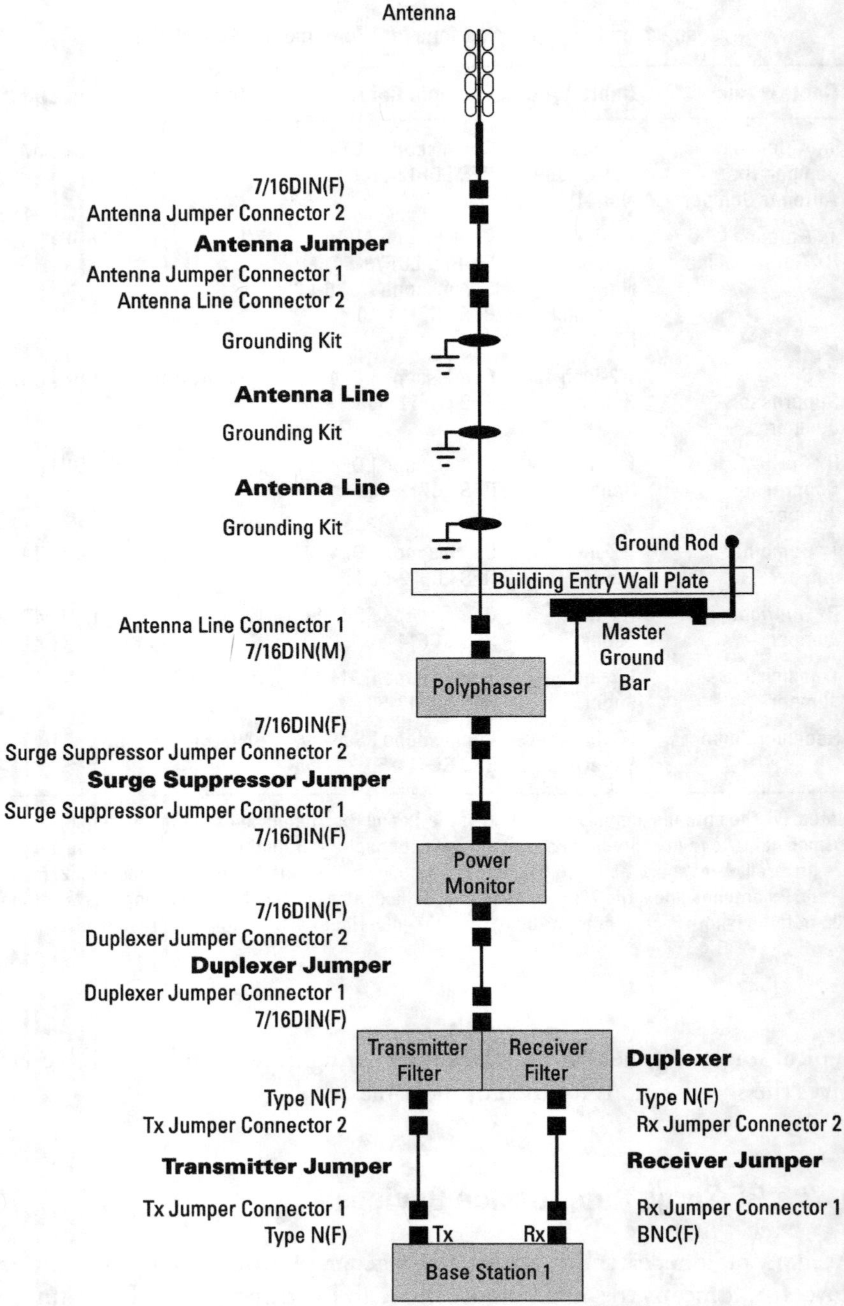

Figure 11.2 Transmission lines for single-channel duplexed site RF system.

Table 11.2
Transmission Line Specifications for Single-Channel Duplexed Site RF System

Cable Name	Cable Type	Example Cables	Connector 1	Connector 2
Antenna Jumper	1/2-inch Heliax (see Note 1)	Commscope LDF4-50, RFS LCF12-50J	7/16DIN (M)	7/16DIN (M)
Antenna Line	7/8-inch Heliax (see Note 2), 1 1/4-inch Heliax	Commscope AVA5-50, RFS LCF78-50 Commscope AVA6-50, RFS LCF114-50	7/16DIN (F)	7/16DIN (F)
Surge Suppressor Jumper	1/2-inch Heliax	Commscope LDF4-50, RFS LCF12-50J	7/16DIN (M)	7/16DIN (M)
Duplexer Jumper	1/2-inch Heliax	Commscope LDF4-50, RFS LCF12-50J	7/16DIN (M)	7/16DIN (M)
Transmitter Jumper	1/2-inch Superflex	Commscope FSJ4-50, RFS SCF12-50	Type N(M)	Type N(M)
Receiver Jumper	1/4-inch Superflex	Commscope FSJ1-50, RFS SCF14-50	BNC (M)	Type N(M)

Notes: (1) The antenna jumper at the end of the antenna line is made from a thinner 1/2-inch cable to relieve pressure on the antenna connector. The antenna jumper is generally 10 ft in length to minimize its effect on system losses. (2) For the antenna line, a 7/8-inch diameter Heliax cable is used if the total length is less than 200 ft. If the length is greater than 200 ft, then a 1 1/4-inch Heliax is used to minimize losses.

and result in costly, large-scale loss of coverage as well as replacement costs. To prevent this, RF surge protection devices are connected in the transmit and receive signal paths. There are several types of RF surge suppression devices, but the basic idea is the same; they pass the RF signal at very low loss but protect the connected RF equipment by almost instantaneously diverting any large surge (such as that caused by a lightning strike) to ground. To make sure the surge energy is grounded well before it could reach the RF equipment, the surge suppression devices are installed at the entry to the equipment shelter, at the junction between the Tx and Rx surge suppressor jumpers and their associated antenna lines as shown in Figure 11.3.

Examples of RF surge suppression devices are the Polyphaser VHF50 (100–512 MHz) and TSX-DFM (700 MHz–2.7 GHz).

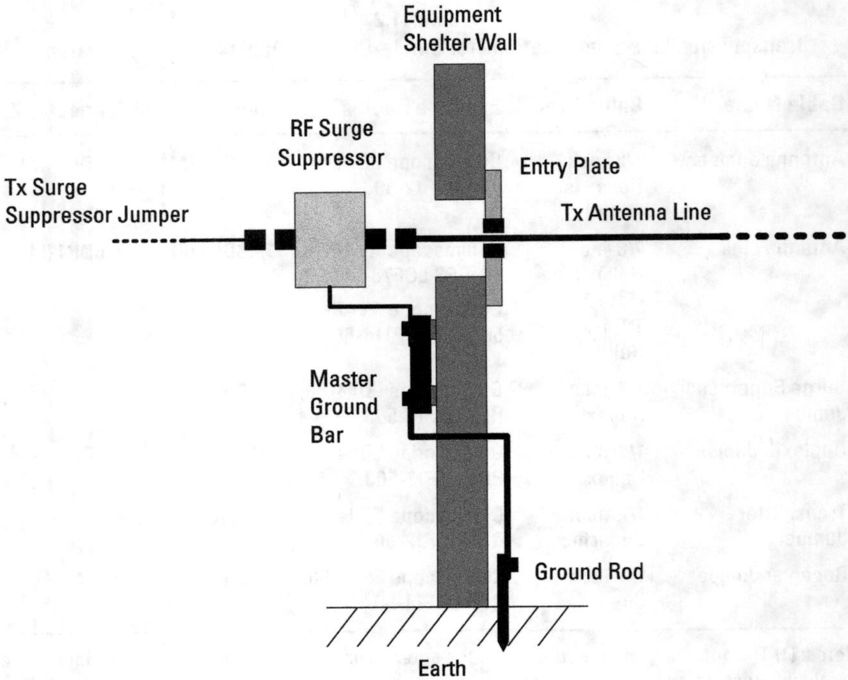

Figure 11.3 Connection of RF surge suppressor.

Just a note here that site grounding is beyond the intended scope of this book and is covered in detail in industry standard documents such as ETSI EG 200 053.

11.4 Transmission Line Grounding Devices

Transmission line grounding devices are effectively clamps that fit around the outer conductor of the Tx and Rx antenna lines and ground them to the tower leg. This ensures that the outer conductor remains at ground potential throughout the length of the Tx and Rx antenna lines, thus minimizing insertion losses. Additionally, in the event of a lightning strike, the grounding devices prevent a large potential difference from developing on the transmission line.

As shown in Figure 11.4, transmission line grounding devices should be installed:

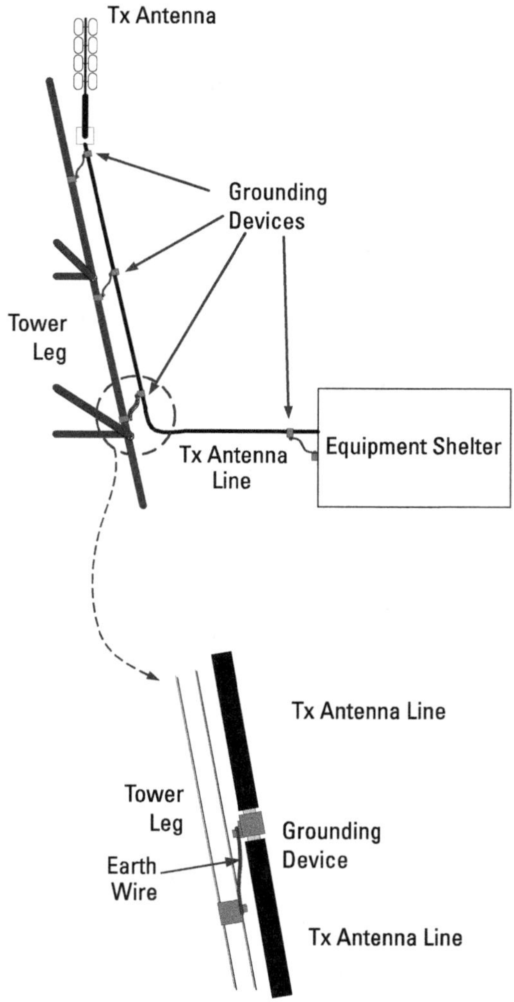

Figure 11.4 Grounding devices on the antenna feeder.

- At the point of exit from the equipment shelter;
- At a point near the bottom of the tower where the Tx or Rx antenna line makes the horizontal to vertical transition;
- At the base of the antenna;
- For towers >200 ft high, in the middle of the Tx or Rx antenna line vertical run up the tower.

All transmission line grounding devices shall be attached to a suitably grounded tower leg or tower ground bus and should be weatherproof to prevent corrosion and subsequent loss of ground contact.

Selected Bibliography

Commscope AVA5-50 data sheet, https://www.commscope.com/globalassets/digizuite/115359-p360-ava5-50-external.pdf.

Commscope AVA6-50 data sheet, https://www.commscope.com/globalassets/digizuite/115393-p360-ava6-50-external.pdf.

Commscope FSJ1-50 data sheet, https://www.commscope.com/globalassets/digizuite/177994-p360-fsj1-50a-external.pdf.

Commscope FSJ4-50 data sheet, https://www.commscope.com/globalassets/digizuite/178000-p360-fsj4-50b-external.pdf.

Commscope LDF4-50 data sheet, https://www.commscope.com/globalassets/digizuite/115611-p360-ldf4-50a-external.pdf.

European Telecommunications Standards Institute, Electromagnetic Compatibility and Radio Spectrum Matters (ERM), Radio Site Engineering for Radio Equipment and Systems, Report ETSI EG 200 053, June 2004.

RFS LCF12-50J data sheet, https://www.rfsworld.com/pim/product/html/LCF12-50J.

RFS LCF78-50JA data sheet, https://www.rfsworld.com/pim/product/html/LCF78-50JA.

RFS LCF114-50JA data sheet, https://www.rfsworld.com/pim/product/html/LCFS114-50JA.

RFS SCF12-50J data sheet, https://www.rfsworld.com/pim/product/html/SCF12-50J.

RFS SCF14-50J data sheet, https://www.rfsworld.com/pim/product/html/SCF14-50J.RFI Technology Solutions TSX-DFF data sheet, https://www.rfi.com.au/core/media/media.nl?id=2633845&c=3389566&h=kotJGklkYWnHBRDK648_G_AJb_7mMI_NRKG5RsmGC0Pd2Zj6&_xt=.pdf.

RFI Technology Solutions VHF50HN data sheet, https://www.rfi.com.au/core/media/media.nl?id=2167221&c=3389566&h=9-J4Dn0SYxjm5lueeHztAXiO_q5x86agEYd1EscQLRCnVEqd&_xt=.pdf.

12

Design Example

Recall the six-site VHF P25 Phase 1 LMR network introduced in the Preface. To recap, this system is intended to provide all the police, fire, and ambulance mission-critical communications for a small city and its surrounding areas.

The purpose of this chapter is to show the whole interference-free site RF antenna and combining system design process for one of the sites in that system, from beginning to end.

To start us off, let us assume that the minimum acceptable delivered audio quality is DAQ3.0 and that, to handle the expected number of calls, each site will have four RF channels.

The design process begins with a provisional frequency plan, which is done by the coverage design engineer to minimize cochannel and adjacent channel interference between the sites. Another thing to avoid is having frequencies that are all equally spaced at any one of the sites, as this will result in many more on-channel IM products being generated.

For our example, let us assume there are 24 VHF frequency pairs assigned to the sites as shown in Figure 12.1. Each frequency pair consists of a transmit and receive frequency with 5-MHz separation between them.

The importance of frequency planning is a fundamental first step towards designing an LMR system that provides maximum coverage at minimum cost.

The remainder of our design example will now focus on designing the antenna and combining system for site 3, which goes by the fictitious name

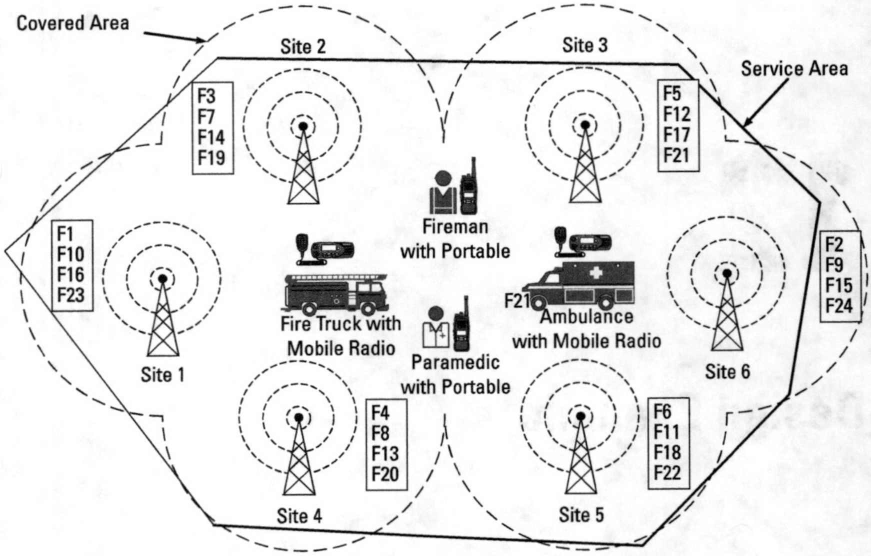

Figure 12.1 LMR design example frequency planning.

of Hilltop Radio Station. The site sits on top of a 500-m-high hill overlooking the city, and Table 12.1 shows that the transmit/receive frequency pairs initially planned for this site are F_5, F_{12}, F_{17}, and F_{21}.

12.1 Information Gathering

As covered in Chapter 2, gathering information so that you have a full understanding of the existing RF assets and environment at the site prior to starting any design work is the most vital step of the entire process.

Table 12.1
Provisional Frequencies for Hilltop Radio Station

Channel	Frequency Pair	Transmit Frequency	Receive Frequency
1	F_5	151.0 MHz	156.0 MHz
2	F_{12}	151.375 MHz	156.375 MHz
3	F_{17}	151.725 MHz	156.725 MHz
4	F_{21}	152.525 MHz	157.525 MHz

Taking the time to uncover the information needed to do the design properly significantly reduces the risk of unexpected interference occurring once the new system is built.

12.1.1 Site Survey

To start with, a site survey was carried out and the site plan shown in Figure 12.2 was drawn up for Hilltop Radio Station. The important information that

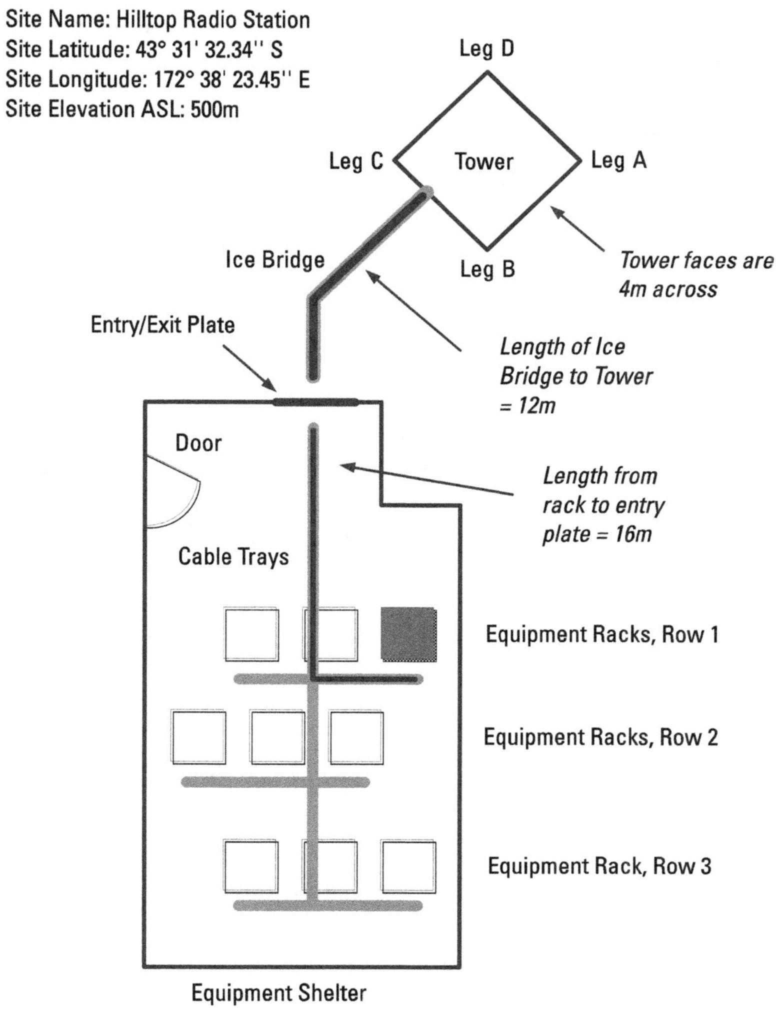

Figure 12.2 Hilltop Radio Station site plan.

this provides is the tower location relative to the shelter, the location of our rack within the shelter, and the lengths of the cable tray and the ice bridge that will carry our RF cables.

12.1.2 Tower Mapping

Tower mapping was carried out next to find out the function of each antenna on the tower of interest and the transmitters or receivers to which they map back in the equipment shelter.

The Hilltop Radio Station tower onto which our system will be deployed is 50m high and each face is 5m wide. Figure 12.3 shows the locations of the existing antennas identified during the survey, and Table 12.2 lists the details of each of these antennas.

Antennas 1a and 1b are the respective receive and transmit antennas for a system operating in the low UHF band. Antennas 2a and 2b are the respective receive and transmit antennas for a system operating in the high UHF band.

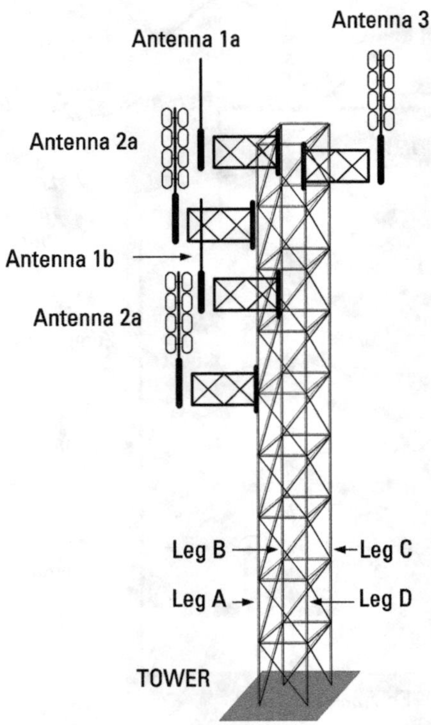

Figure 12.3 Hilltop Radio Station tower mapping.

Table 12.2
Hilltop Radio Station Tower Mapping

Antenna	Leg	Base Height	Antenna	Azimuth	Tx Channels	Tx EIRP	Rx Channels
1a	B	49m	Colinear	180°	NA	NA	430.5 MHz, 431.5 MHz
1b	B	43m	Colinear	180°	421.5 MHz, 422.5 MHz	+47.8 dBm	NA
2a	A	45m	Exposed dipole	90°	NA	NA	480.5 MHz, 481.5 MHz, 482.8 MHz, 483.6 MHz
2b	A	39m	Exposed dipole	90°	471.5 MHz, 472.5 MHz, 473.8 MHz, 474.6 MHz	+49.1 dBm	NA
3	D	49m	Exposed dipole	0°	136.0 MHz	+52.0 dBm	140.6 MHz

As the system that we are planning to deploy operates in the low VHF band, neither of these antennas are of concern as they are in different bands where the frequency separation alone will be enough to prevent interference. Antenna 3 is a duplexed low VHF band antenna to which we will need to pay close attention to when planning where to locate our transmit and receive antennas.

12.1.3 RF Survey

Being conveniently located at a geographic high point means Hilltop Radio Station is a popular location for radio system planners, so there is more than one radio tower there as illustrated in Figure 12.4.

Mapping all the towers and then working out the levels of all the interfering signals our system is going to be exposed to would be very time-consuming. So, instead, an RF survey of 24 hours duration was carried out, starting with signal monitoring to pick up any existing RF activity that may affect our proposed new design for this site. The monitored band was 141 to 173 MHz, as this represents our span of receive frequencies (156.0 to 157.525 MHz) plus an extra 10% above and below.

The test setup for signal monitoring shown in Figure 12.5 uses a wideband antenna placed close to the tower, followed by a 141 to 173-MHz bandpass

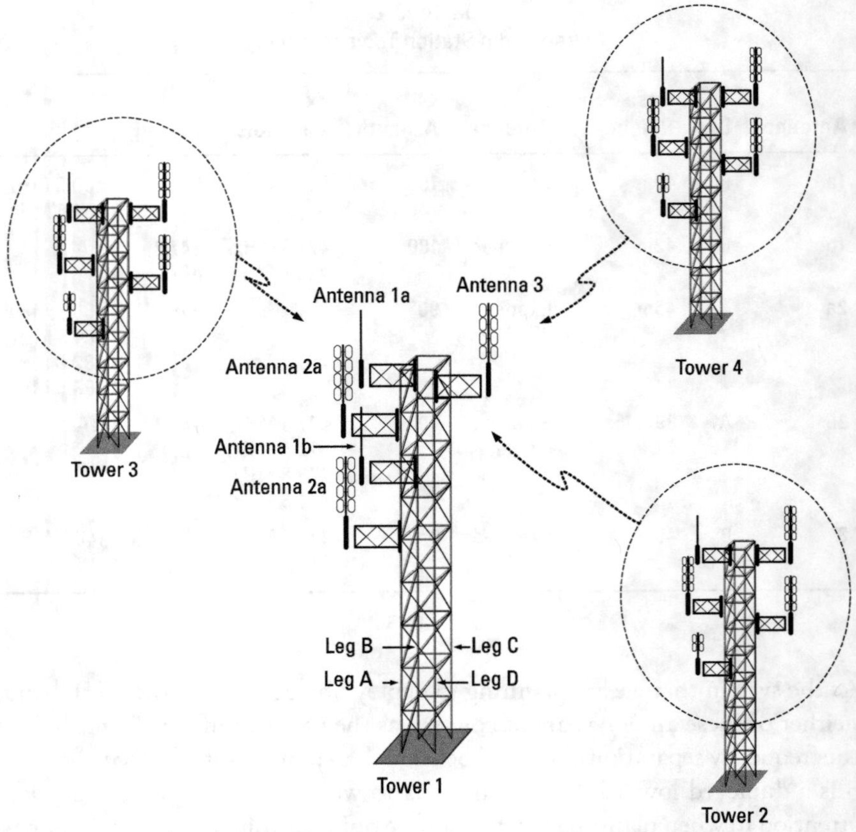

Figure 12.4 Interference from colocated towers at Hilltop Radio Station.

filter and a spectrum analyzer that measures the frequency and level of any existing RF signals picked up within the band.

As described in Chapter 2, it is the unwanted high-level carriers (HLCs) exceeding –40 dBm that are present more than 5% of time that have the potential to contribute towards interference that will adversely impact the system performance. Table 12.3 shows four in-band HLCs picked during the signal monitoring.

The first three HLCs must be considered in IM studies and may require specific filtering within our new combining system design. The fourth HLC (168.6 MHz) can be disregarded as it was only present 1.2% of the time and is unlikely to contribute to interference that adversely affects our system.

Next, noise monitoring was carried out to measure the ambient site noise level over the entire receive band being considered for our new system.

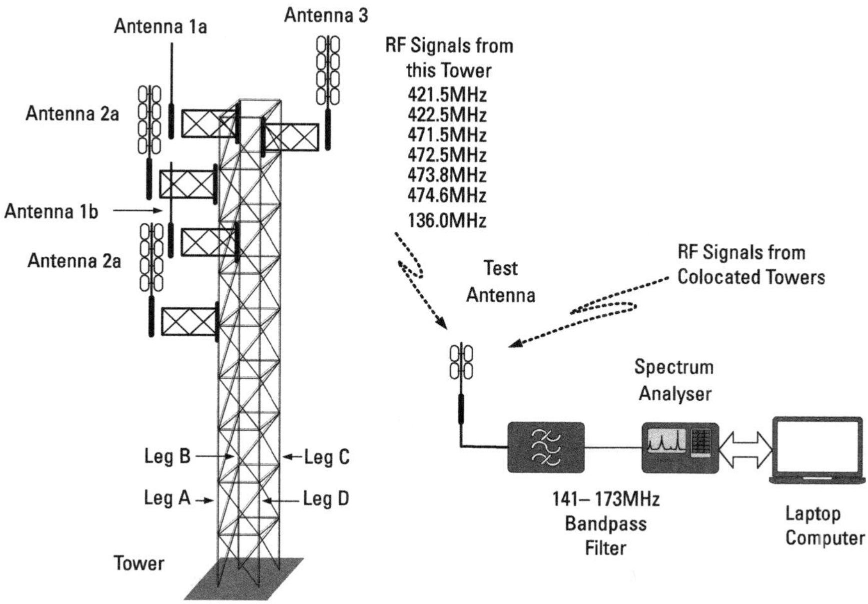

Figure 12.5 Signal monitoring at Hilltop Radio Station.

A frequency agile crystal filter and a low noise preamplifier were added to the test system as shown in Figure 12.6 to allow measurement of noise levels very close to thermal.

The measured results giving the ambient site noise level observed at each 12.5-kHz LMR channel across the receive band are shown in Table 12.4. The thermal noise generated in the test system because of its noise figure has been subtracted from these measurements so all we have left is the actual site noise. Giving the site noise levels in dB above KTB means the measurements are

Table 12.3
Signal Monitoring Results Showing HLCs

Frequency	Average Power	Occupancy
146.0 MHz	–8 dBm	15%
161.875 MHz	–22 dBm	27%
166.225 MHz	–27 dBm	100%
168.6 MHz	–24 dBm	1.2%

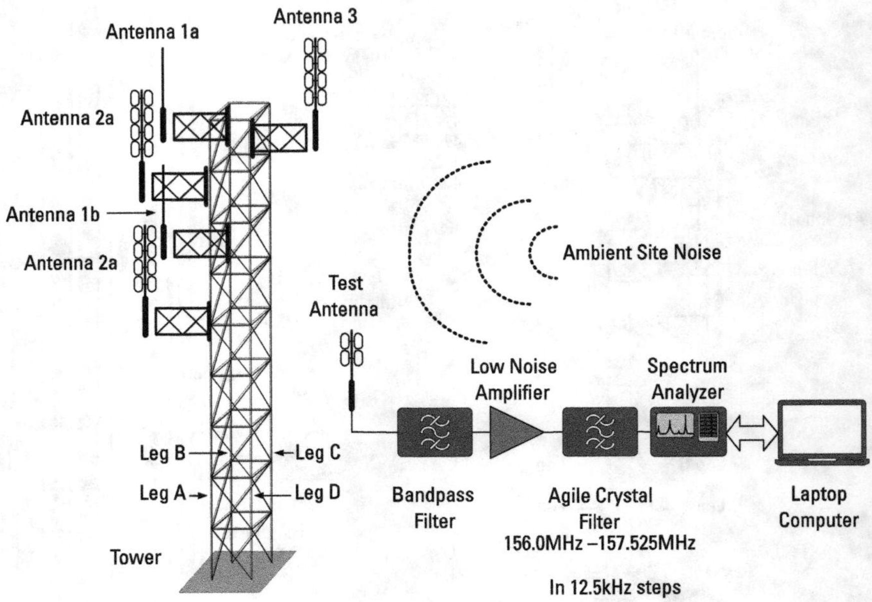

Figure 12.6 Noise monitoring at Hilltop Radio Station.

bandwidth agnostic and so can be directly used even if the noise bandwidth of the test system and the LMR receivers are different.

At this point, it is worth showing our planned transmit and receive channels in Figure 12.7 alongside the interfering signals identified during the tower mapping and RF survey. A diagram like this gives an easy appreciation of the RF environment our system will be operating within.

Having now gathered all the information necessary, let us move on to see how we use it to design the full interference-free site RF antenna and combining system for the Hilltop Radio Station site.

12.2 Antenna System Design

The first step in the design is choosing where to locate our antennas on the Hilltop Radio Station tower.

The rules for designing optimal antenna systems for deployment on radio towers were covered in Chapter 5. Applying the design rules, in turn, then to develop the Hilltop antenna system, our site RF antenna and combining system is multichannel, so rule 1 states that we should use separate transmit and receive antennas, as shown in Figure 12.8.

Table 12.4
Noise Monitoring Results Showing Site Noise Floor above KTB in Each Channel

Frequency	Noise above Thermal (KTB)
156.0	12.0
156.0125	12.3
156.025	12.2
↓	↓
156.3625	12.7
156.375	12.9
156.3875	12.9
↓	↓
156.7125	13.4
156.725	13.4
156.7375	13.7
↓	↓
157.5	14.2
157.5125	14.4
157.525	14.4

Now we need to apply rules 2 to 5 to decide on a location for our receive antenna. To maximize uplink (talk-in) coverage because that is usually the weakest path, the receive antenna is placed uppermost. Ideally, this would be right at the top of the tower, but, in this case, there is a VHF duplexed antenna already there (antenna 3 from the tower mapping). To mitigate potential interference, there needs to be 60 dB of isolation between our receive antenna and antenna 3, which effectively means that our receive antenna has to be located lower down the tower. The next thing is to choose a leg on which to locate the receive antenna. Looking at Figure 12.1, Hilltop Radio Station (site 3) is located in the northeast part of the service area, so, to optimize coverage, the best legs to locate our antennas on would be either legs B or C (see Figure 12.2). However, leg B already has two existing UHF antennas (antennas 1a and 1b) located there, meaning that our receive antenna would also have to be below those. So, the logical choice of the leg on which to place our antenna for best coverage into the service area is leg C. Having decided that, we now need to work out the highest elevation that we can mount our receive antenna at on leg C so that it maintains 60 dB of isolation from antenna 3 at 49m on

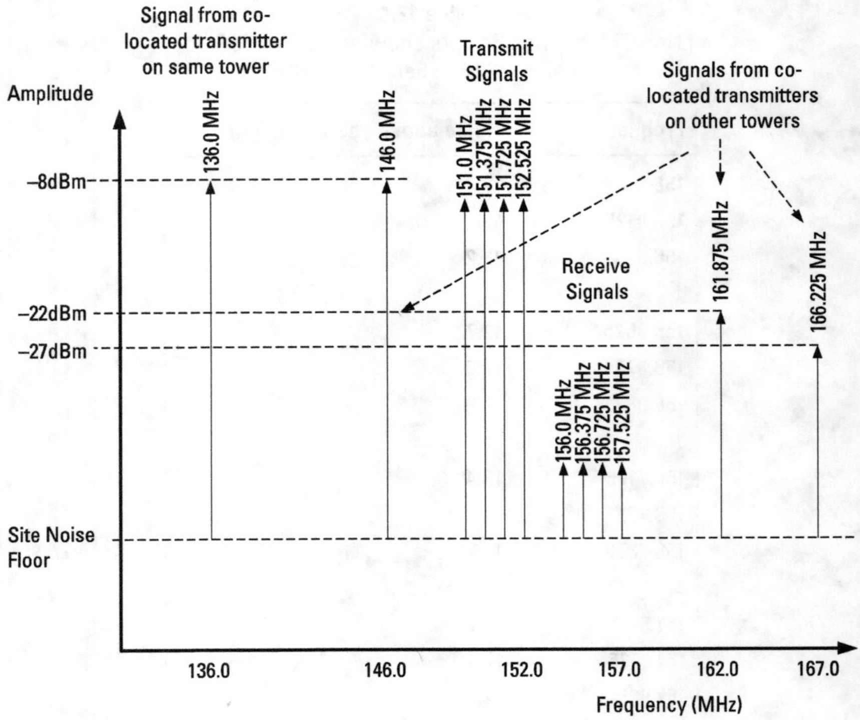

Figure 12.7 Design example RF environment.

leg D. Because our receive antenna and antenna 3 will be on different legs, the separation between them is a combination of both horizontal and vertical separation (known as slant separation), as shown in Figure 12.9.

So, let us start with the equation for calculating the isolation between slant separated antennas.

$$I_S = (I_V - I_H) \times (\alpha/90) + I_H \qquad (12.1)$$

where

I_S = slant isolation (dB);

I_H = horizontal isolation (dB);

I_V = vertical isolation (dB);

α = vertical angle between Tx and Rx antennas (degrees).

The easiest part to calculate to begin with is the horizontal isolation, I_H. To do this, antenna 3 is mounted on a 0.9-m standoff in accordance with

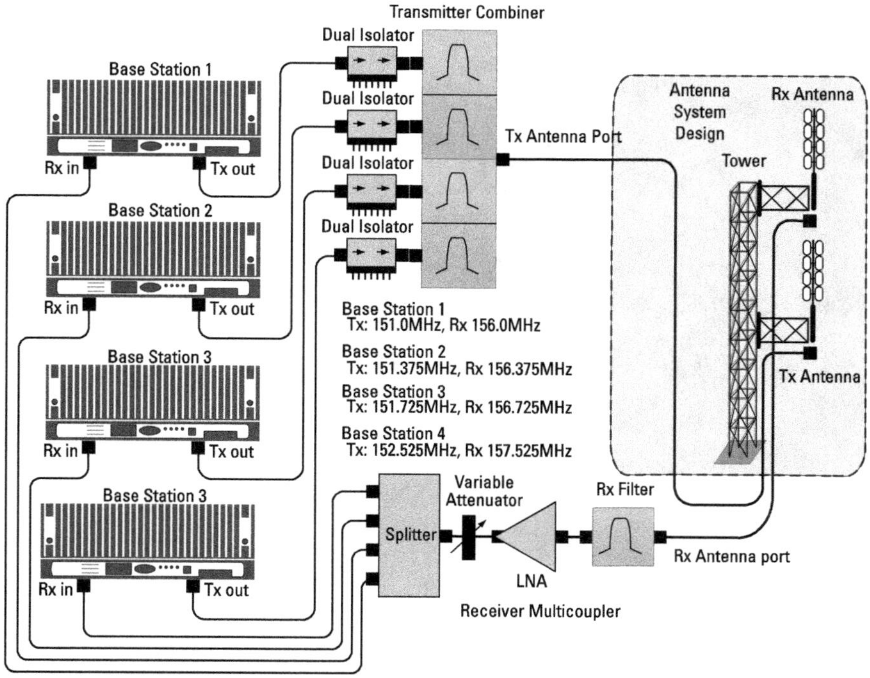

Figure 12.8 Hilltop Radio Station RF system with separate transmit and receive antennas.

antenna system design rule 8, and our new receive antenna mounted on a 1.8-m standoff (as it cannot be at the top of the tower). If each standoff points out at 45° from the tower leg, then the horizontal separation between them by simple trigonometry is 6.9m. The tower mapping drawing shows that antenna 3 is a 4-stack exposed dipole with a gain of 8.15 dBi. To maximize coverage, because Hilltop Radio Station is situated near the northeast corner of the service area, a 4-stack offset array antenna would be a good choice for our system as it will radiate more favorably in towards the area that we need to cover.

With this, we can now calculate the horizontal isolation, I_H; thus:

$$I_H = 22 + 20 \times \mathrm{LOG}_{10}\ (d_h / \lambda) - (G_{\mathrm{TX}} + G_{\mathrm{RX}}) \tag{12.2}$$

where

I_H = horizontal isolation (dB);

d_h = horizontal separation (m);

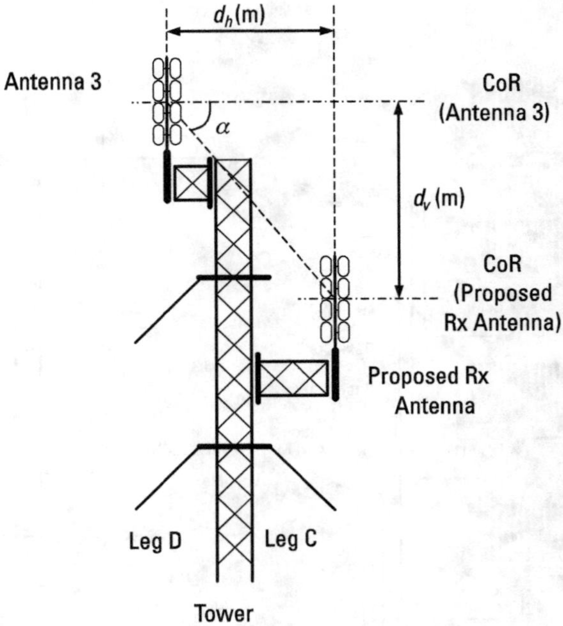

Figure 12.9 Slant separated antennas.

λ = wavelength (m) = 1.98m at the average of our operating frequencies;

G_{TX} = gain of Tx antenna (dBi);

G_{RX} = gain of Rx antenna (dBi).

So

$$\begin{aligned} I_H &= 22 + 20 \times \mathrm{LOG}_{10}\,(d_h/\lambda) - (G_{TX} + G_{RX}) \\ &= 22 + 20 \times \mathrm{LOG}_{10}\,(6.9/1.98) - (8.15 + 8.15) \\ &= 16.9 \text{ dB} \end{aligned} \tag{12.3}$$

Going back to the formula for slant isolation, we now know I_H, but we still need to find I_V and α. The trouble is that they are dependent on each other; we cannot just work out the vertical separation that we need because in doing so the angle α between centers of radiation of antenna 3 and our receive antenna changes as well. The simplest thing to do is substitute an educated guess value for the vertical isolation, I_V, into the slant isolation equation and see what happens. The equation for slant isolation tells us that $I_V > I_S$, so if we need a slant isolation of 60 dB, then I_V must be greater, say, 70 dB. So, let us now transpose the equation for vertical isolation I_V to see what the required physical separation for 70-dB vertical isolation is.

$$I_V = 28 + 40 \times \mathrm{LOG}_{10}\,(d_v / \lambda) \tag{12.4}$$

where

I_V = vertical isolation (dB);

d_v = vertical separation (m);

λ = wavelength (m).

Hence,

$$\begin{aligned} d_v &= \lambda \times 10^{[(I_v - 28)/40]} \\ &= 2 \times 10^{[(70 - 28)/40]} \\ &= 22.4\text{m} \end{aligned} \tag{12.5}$$

Knowing now that the horizontal separation between antenna 3 on leg D and our receive antenna on leg C is 6.9m, and the vertical separation is 22.4m, it is now a simple matter to calculate the angle between the centers of radiation, α, shown in Figure 12.10.

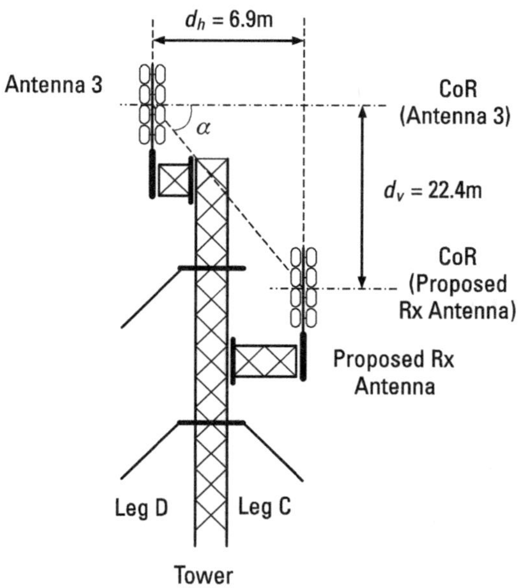

Figure 12.10 Angle between slant separated antennas.

To calculate the angle between the centers of radiation, α:

$$\begin{aligned}\text{Tan } \alpha &= (22.4/6.9) \\ &= 3.25\end{aligned} \tag{12.6a}$$

So

$$\begin{aligned}\alpha &= \tan^{-1}(3.25) \\ &= 72.9^\circ\end{aligned} \tag{12.6b}$$

Now let us go back and see if our educated guess resulted in the 60 dB of slant isolation that we are looking for:

$$I_S = (I_V - I_H) \times (\alpha/90) + I_H \tag{12.7a}$$

where

I_S = slant isolation (dB);
I_H = horizontal isolation (dB);
I_V = vertical isolation (dB);
α = vertical angle between Tx and Rx antennas (degrees).

So:

$$\begin{aligned}I_S &= (70 - 16.9) \times (72.9/90) + 16.9 \\ &= 60.0 \text{ dB}\end{aligned} \tag{12.7b}$$

So, our receive antenna on leg C is 22.4m below the height of antenna 3 on leg D, which makes it 26.6m above ground level as shown in Figure 12.11.

The next step is to locate the transmit antenna. To maximize isolation, this needs to be placed directly below the receive antenna on leg C. Antenna system design rule 3 (see Chapter 5) states that, for maximum coverage and minimum interference potential and to optimize transmitter combiner and receiver multicoupler size and cost, we need at least 60 dB of isolation between the transmit and receive antennas. So, let us transpose the equation for vertical isolation I_V to see what the required physical separation is.

$$I_V = 28 + 40 \times \text{LOG}_{10}(d_v/\lambda) \tag{12.8}$$

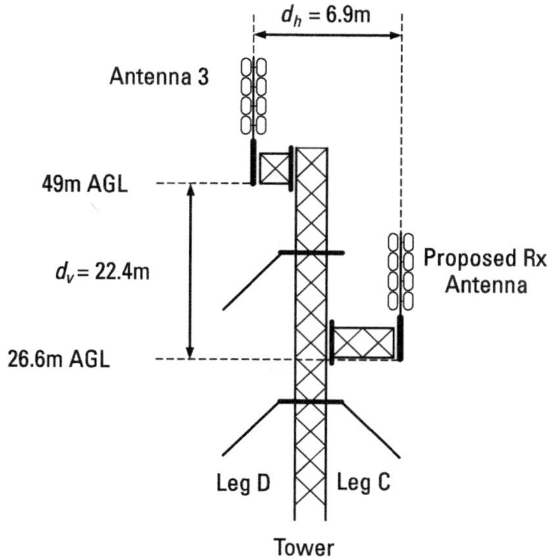

Figure 12.11 Receive antenna position.

where

I_V = vertical isolation (dB);

d_v = vertical separation (m);

λ = wavelength (m).

Hence,

$$
\begin{aligned}
d_v &= \lambda \times 10^{[(I_v - 28)/40]} \\
&= 1.98 \times 10^{[(60 - 28)/40]} \qquad (12.9) \\
&= 12.5\text{m}
\end{aligned}
$$

This places our transmit antenna on leg C at 14.1m above ground level (which is 12.5m below our receive antenna on leg C).

The antenna design for our new system is now complete; it complies with the design rules and resulted in the locations shown in Figure 12.12.

These heights might seem exceedingly low, but as the Hilltop tower is situated on a hill 500m high, the actual effect on coverage from having our antennas relatively low on the tower is minimal.

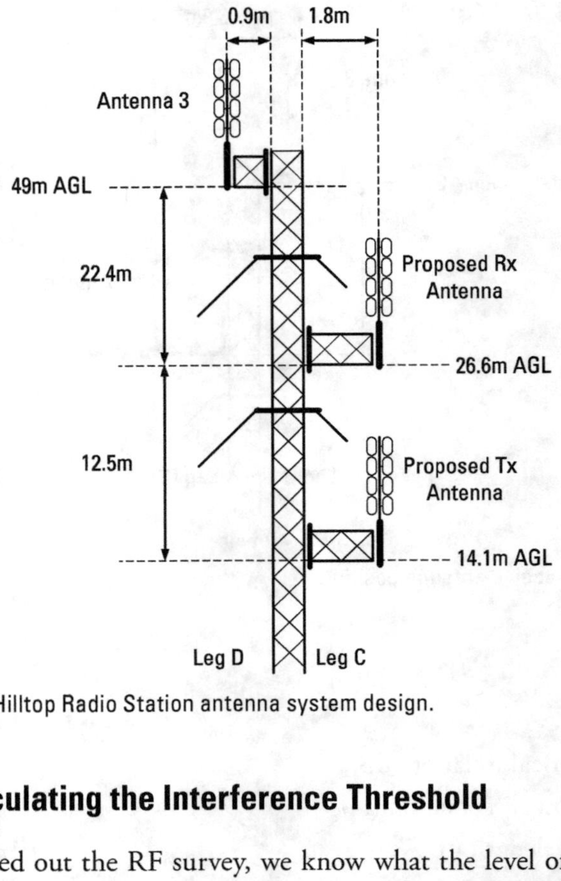

Figure 12.12 Hilltop Radio Station antenna system design.

12.3 Calculating the Interference Threshold

Having carried out the RF survey, we know what the level of ambient site noise is at the Hilltop Radio Station site receive frequencies (see Table 12.4). This site noise places a direct limitation on the receiver sensitivity that our new LMR system can achieve and hence the distance from which a user making a call can be heard. In addition to the existing site noise, when we install our LMR system at the site, its transmitters will add some extra noise and interference of their own, which can never be zero. So, what we need to do now is calculate what level of extra noise and interference that we are willing to let our system add to the existing site noise floor. Generally, what is considered acceptable is a level of extra noise and interference that causes the existing site noise floor to rise by 1 dB. Put another way, the level of extra noise is that which causes the sensitivity of our receivers to degrade by 1 dB compared to what it would have been with the existing site noise floor. This will only have a minimal effect on coverage.

As our system is a P25 Phase 1 system using C4FM modulation designed for a minimum acceptable delivered audio quality (DAQ) of 3.0, then the received signal must be at least 15.2 dB above the sum of noise and interference (see Chapter 6, Table 6.1).

As covered in Chapter 6, there are several steps involved in calculating the interference threshold, starting by working out the thermal noise floor of the receiving system (consisting of the receiver filter, receiver multicoupler, and the connected receivers).

$$N_O\ (\text{dBm}) = -174 + [10 \times \text{LOG}_{10}\ (B)] + Rx\ NF \qquad (12.10\text{a})$$

where

N_O = thermal noise power referred to the base of the receive antenna (dBm);

B = filter pass bandwidth of the susceptible receiver (Hz);

$Rx\ NF$ = noise figure of the susceptible receiving system (including antenna cabling, filters, multicoupler, and base station receiver) (dB).

As practical LMR receivers have a filter pass bandwidth of around 8 kHz and a noise figure of around 8 dB, we get a thermal noise floor of:

$$\begin{aligned} N_O\ (\text{dBm}) &= -174 + [10 \times \text{LOG}_{10}\ (8 \times 10^3)] + 8 \\ &= -127\ \text{dBm/8 kHz} \end{aligned} \qquad (12.10\text{b})$$

The next step to calculate the best possible receiver faded sensitivity, which is basically the minimum signal level required at the base of the receive antenna to provide a demodulated audio quality of DAQ3.0 in a Rayleigh faded environment. We do this by adding the $C_F\ /(N + I)$ threshold to the previously calculated thermal noise floor (N_O).

$$\begin{aligned} &\text{Best possible dynamic receiver sensitivity (dBm)} \\ &= N_O + [C_F\ /(N + I)] \\ &= -127 + 15.2 \\ &= -111.8\ \text{dBm} \end{aligned} \qquad (12.11)$$

where

Dynamic receiver sensitivity = the received signal level for minimum performance (dBm);

N_O = thermal noise power referred to the base of the receive antenna (dBm);

$C_F/(N+I)$ = the Rayleigh faded carrier-to-noise and interference ratio (dB).

This is the best the receiver sensitivity can possibly be for the Hilltop site because the calculation does not yet include any ambient site noise or external interference. To incorporate the effect of site noise, we refer to the noise floor results measured during the RF survey (see Table 12.4) and look up the worst site noise level on any of the receive frequencies, which was 14.4 dB above thermal at 157.525 MHz. So, to convert this noise floor to an actual level:

$$N_S\ (\text{dBm}) = -174 + [10 \times \text{LOG}_{10}\ (B)] + S \qquad (12.12a)$$

where

N_S = site noise referred to the base of the receive antenna (dBm);

B = filter pass bandwidth of the susceptible receiver (Hz);

S = site noise (dB above thermal noise).

So, in this example:

$$\begin{aligned} N_S\ (\text{dBm}) &= -174 + [10 \times \text{LOG}_{10}\ (B)] + S \\ &= -174 + [10 \times \text{LOG}_{10}\ (8 \times 10^3)]] + 14.4 \qquad (12.12b) \\ &= -120.6\ \text{dBm} \end{aligned}$$

The next step is to find the total noise (N_T) in the receiver bandwidth by adding the site noise (N_S) to the previously calculated thermal noise (N_O), that is,

$$N_T = 10 \times \text{LOG}_{10}\ (N_O + N_S) \qquad (12.13a)$$

where

N_T = total noise referred to the base of the receive antenna (dBm);

N_S = site noise referred to the base of the receive antenna (mW);

N_O = thermal noise power referred to the base of the receive antenna (mW).

Note that the site noise (N_S) and thermal noise (N_O) both need to be converted to milliwatts first before they can be summed together.

To get back to the calculation, then:

$$\begin{aligned} N_O\ (\text{mW}) &= 10^{(N_O\ (\text{dBm})/10)} \\ &= 10^{(-127/10)} \\ &= 1.99 \times 10^{-13}\ \text{mW} \end{aligned} \tag{12.13b}$$

and

$$\begin{aligned} N_S\ (\text{mW}) &= 10^{(N_S\ (\text{dBm})/10)} \\ &= 10^{(-120.6/10)} \\ &= 8.77 \times 10^{-13}\ \text{mW} \end{aligned} \tag{12.13c}$$

So:

$$\begin{aligned} N_T &= 10 \times \text{LOG}_{10}\ (N_O + N_S) \\ &= 10 \times \text{LOG}_{10}\ [(1.99 \times 10^{-13}) + (8.77 \times 10^{-13})] \\ &= -119.7\ \text{dBm} \end{aligned} \tag{12.13d}$$

To this, we add 1-dB allowable receiver desensitization, so the total noise floor (N_x) becomes:

$$N_X = N_T + 1\ \text{dB} \tag{12.14a}$$

where

N_X = total noise and interference floor referred to the base of the receive antenna (dBm);

N_T = total noise referred to the base of the receive antenna (dBm).

So

$$\begin{aligned} N_X &= N_T + 1\ \text{dB} \\ &= -119.7 + 1 \\ &= -118.7\ \text{dBm} \end{aligned} \tag{12.14b}$$

Now, we can calculate the dynamic receiver sensitivity in the presence of site noise by adding the relevant $C_F/(N + I)$ ratio to the new total noise and interference floor (N_X), that is,

$$\begin{aligned} \text{Dynamic receiver sensitivity (dBm)} &= N_X + [C_F/(N + I)] \\ &= -118.7 + 15.2 \\ &= -103.5\ \text{dBm} \end{aligned} \tag{12.15}$$

where

Dynamic receiver sensitivity = the received signal level required in the presence of site noise for minimum performance (dBm);

N_X = total noise and interference floor referred to the base of the receive antenna (dBm);

$C_F/(N+I)$ = the Rayleigh faded carrier to-noise and interference ratio (dB).

Now the next step is to work out the interference threshold (I_T), the level of interference that adds to the total noise floor (N_T) and causes the receiver sensitivity to degrade to −103.5 dBm.

$$I_T = 10 \times \text{LOG}_{10}\,(N_X - N_T) \tag{12.16a}$$

where

I_T = interference threshold, the total interference level referred to the base of the receive antenna (dBm);

N_X = total noise and interference floor referred to the base of the receive antenna (mW);

N_T = total noise floor referred to the base of the receive antenna (mW).

We previously calculated $N_T = -119.7$ dBm and $N_X = -118.7$ dBm, so, converting to milliwatts:

$$\begin{aligned} N_X\,(\text{mW}) &= 10^{(N_X\,(\text{dBm})/10)} \\ &= 10^{(-118.7/10)} \\ &= 1.35 \times 10^{-12}\ \text{mW} \end{aligned} \tag{12.16b}$$

and

$$\begin{aligned} N_T\,(\text{mW}) &= 10^{(N_T\,(\text{dBm})/10)} \\ &= 10^{(-119.7/10)} \\ &= 1.07 \times 10^{-12}\ \text{mW} \end{aligned} \tag{12.16c}$$

So, by substituting in these values, we find the interference level (I_T):

$$\begin{aligned} I_T &= 10 \times \text{LOG}_{10}\,(N_X - N_T) \\ &= 10 \times \text{LOG}_{10}\,[(1.35 \times 10^{-12}) - (1.07 \times 10^{-12})] \\ &= -125.5\ \text{dBm} \end{aligned} \tag{12.16d}$$

That is the total level of interference that we can tolerate at the base of the receive antenna if we are only to suffer 1 dB of receiver desensitization.

The final part of the problem is there can be multiple interference mechanisms at play. As described in Chapter 4, local interference at a shared radio site can be in the form of:

- Harmonics from other transmitters;
- Transmitter IM where products generated within a transmitter combiner are radiated at the same frequency to which a nearby receiver is tuned;
- Transmitter sideband noise from colocated transmitters that falls inside the susceptible receiver's pass bandwidth;
- Receiver blocking where a very strong off-channel signal from a nearby transmitter overloads the susceptible receiver;
- Receiver IM produced in the site receiving system as a result of all the interfering signals that appear on the receive antenna.

While it is unlikely that all these mechanisms would occur simultaneously at a given shared radio site, it is probable that more than one mechanism is having an influence. If only one of these mechanisms is occurring, then the interference per mechanism is the same as the total interference level, I_T. If two of the five mechanisms are present, then the interference per mechanism is equal to $I_T - 3$ dB, … . We do not know how many of these interference mechanisms are present, as we have yet to do the IM analysis. Once we have done this during the transmitter combiner and receiver multicoupler design calculations, we will need to revise the interference threshold based on the number of interference mechanisms that are actually affecting our receivers. For now, representing the calculation on Figure 12.13, this is what we have.

12.4 Transmitter Combiner Design

Now that the positions of the antennas on the Hilltop Radio Station tower have been finalized and the interference threshold calculated, the next step is to design the transmitter combining system shown in Figure 12.14 according to the design rules given in Chapter 8 so that it achieves the goals of minimum interference potential and hence maximum coverage.

Checking through some of the initial transmitter combiner design rules, we find rule 1 just reemphasizes the importance of using separate transmit and receive antennas for multichannel repeater site combining systems, as we have

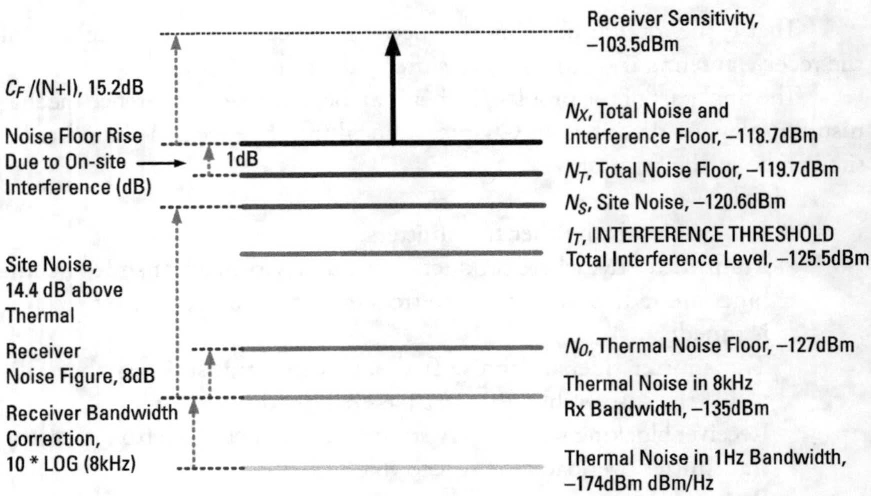

Figure 12.13 Calculation of interference threshold for Hilltop Radio Station.

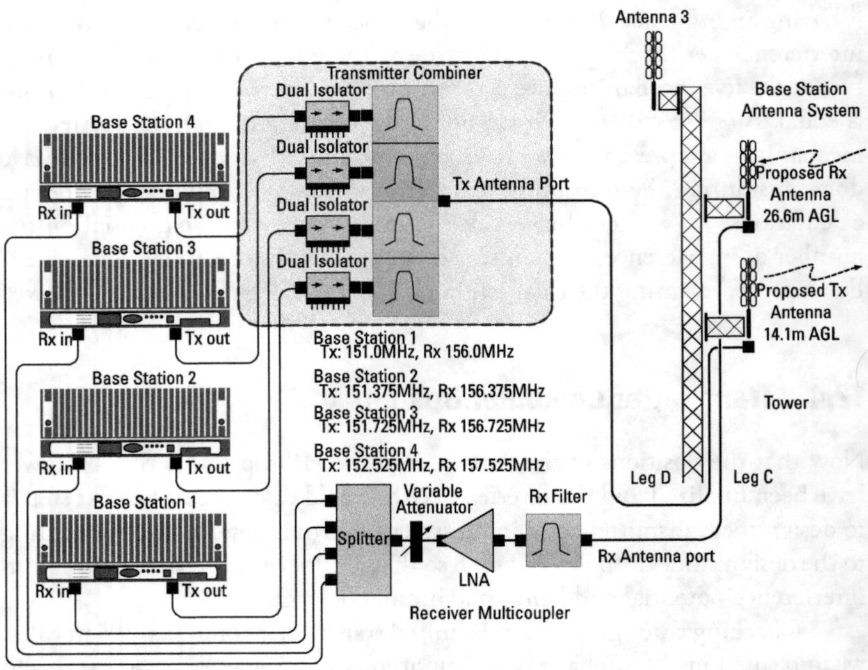

Figure 12.14 Four-channel transmitter combining system for the Hilltop Radio Station.

already done within the antenna system design part of this design example. We also comply with rule 2, as we are only combining four transmitters onto one transmit antenna. To comply with rule 4 (we will come back to rule 3 as it is related to rule 5), we will design our transmitter combiner using highly selective bandpass cavity filters. Each cavity filter will be tuned to the transmit frequency of the base station to which it is connected, with the outputs then connected in a T-Pass fashion with coaxial cables. T-Pass combiners are preferable over Star combiners, as they are easier to add more channels to later. Rule 3 states the insertion loss of the combiner shall be no more than 4 dB in any of the transmit channels. Meeting this goal is partly a function of the frequency separation between the transmitters connected to the combiner, which are given in Table 12.5.

As we can see, this easily complies with the 75-kHz minimum recommended transmit-to-transmit frequency separation for a system operating in the low VHF band, so we should have no problem meeting the 4.0-dB insertion loss goal.

As well as combining four transmitters on to one antenna to save tower real estate, our transmitter combiner performs two more important functions in managing interference:

- Mitigation of transmitter IM;
- Suppressing transmitter sideband noise at all site receive frequencies.

Considering transmitter IM first, this is IM generated within the transmitter combining and antenna system whereby the transmitter signals mix in any nonlinearities.

For systems like ours where we have 60 dB of isolation between the transmit and receive antennas, design rule 7 tells us that we need to search for third, fifth, and seventh-order transmitter IM products within a search bandwidth of:

Table 12.5
Transmitter Frequency Separation for Hilltop Radio Station

Channel	Transmit Frequency	Separation to Nearest Transmitter
1	151.0 MHz	375 kHz
2	151.375 MHz	350 kHz
3	151.725 MHz	350 kHz
4	152.525 MHz	800 kHz

$$\text{Search bandwidth (kHz)} = (n + 1) \times \text{Channel bandwidth} \quad (12.17)$$

where

n = maximum order of IM to be searched for;

Channel bandwidth = 12.5 kHz (typical for LMR systems).

So, for a maximum of seventh-order IM and a 12.5-kHz channel bandwidth, the search bandwidth is 100 kHz (+/−50 kHz around each receive frequency). This wider search bandwidth is necessary because the occupied bandwidth of an IM product spreads according to its order, so while the center of the IM product may not fall directly on a receive channel, some of the energy can spill over into a receiver's passband.

In summary then, we need to see whether any third, fifth, and seventh-order combinations of the transmit frequencies in Table 12.6 fall within +/−50 kHz of any of the receive frequencies.

Having run the analysis, Table 12.7 shows there are a few seventh-order IM products that fall within the search bandwidth of the receive channels.

Note here that a similar transmitter IM analysis would be carried out at any other VHF receive frequencies at the site too, as the idea is that we should not emit transmitter IM products that fall on any site receive frequency.

To mitigate the seventh-order IM products identified, we take the following steps in the transmitter combiner design:

- Specify that 80-dB of isolation must exist between the individual transmitter outputs on the combiner. This means that we must use dual isolators in the outputs of each transmitter. To comply with design rule 8, the dual isolator dump load nearest the cavity filters will need to be capable of handling the full power from the connected transmitter in the event of a fault, as 100% reflection could occur in extreme cases.

Table 12.6
Frequencies for Hilltop Radio Station Transmitter IM Analysis

Channel	Frequency Pair	Transmit Frequency	Receive Frequency
1	F_5	151.0 MHz	156.0 MHz
2	F_{12}	151.375 MHz	156.375 MHz
3	F_{17}	151.725 MHz	156.725 MHz
4	F_{21}	152.525 MHz	157.525 MHz

Table 12.7
Hilltop Radio Station Transmitter IM Results

	Number of Transmitter IM Products at Each Receive Frequency							
	156.0 MHz		156.375 MHz		156.725 MHz		157.525 MHz	
IM Order	**Direct**	**Indirect**	**Direct**	**Indirect**	**Direct**	**Indirect**	**Direct**	**Indirect**
Third	0	0	0	0	0	0	0	0
Fifth	0	0	0	0	0	0	0	0
Seventh	1	2	1	1	1	0	0	0

- Use a durable 1/2-inch-diameter Heliax cable for the transmission lines connecting the transmitter combiner output to the transmit surge suppressor input.
- Use a durable 7/8-inch-diameter Heliax cable for the transmission line connecting the transmit surge suppressor output to the transmit antenna.
- Use a 7/16-inch DIN connectors everywhere between the output of the transmitter combiner to the transmit antenna.
- Use a PIM-rated transmit antenna that has all connections welded, not crimped.

The next step of the design is to consider transmitter sideband noise. Design rule 9 states that we need to calculate and specify the sideband noise suppression required in each arm of the transmitter combiner so no more than 1 dB of desensitization of any site receiver results. To do this, we need to know the output power of each transmitter at the transmit antenna port, their individual transmitter sideband noise levels at the nearest receive frequency, the isolation between the transmit and receive antennas, and the interference threshold.

To start with, we already know that:

- Interference threshold = −125.5 dBm;
- Transmit to receive antenna isolation = 60 dB;
- Receiver equivalent noise bandwidth, $B = 8$ kHz.

In addition, we are combining four transmitters, and let us assume:

- Transmitter power (per channel at the transmit antenna) P_{TX} = +45 dBm. This is the power after the transmitter combiner loss (3 dB) and the RF cabling loss (2 dB);
- Transmitter sideband noise from each transmitter at the nearest receive frequency, $Tx\ SBN_O = -147$ dBc/Hz.

With this information, we can start by working out the total transmitter sideband noise radiated by the transmit antenna at the nearest receive frequency.

$$Tx\ SBN\ (\text{dBm/8 kHz}) = Tx\ SBN_O\ (\text{dBc/Hz}) + P_{TX} + [10 \times \text{LOG}_{10}\ (B)] + [10 \times \text{LOG}_{10}\ (Tx)] \quad (12.18a)$$

where

$Tx\ SBN$ = transmitter sideband noise that falls in the receiver's bandwidth (dBm/8 kHz);

$Tx\ SBN_O$ = normalized transmitter sideband noise at the receive frequency (dBc/Hz);

P_{TX} = output power per transmitter at the transmit antenna output (dBm);

B = filter pass bandwidth of the susceptible receiver (Hz);

Tx = number of transmitters on the combiner.

So, in this case:

$Tx\ SBN$ (dBm/8 kHz)

$$\begin{aligned} &= Tx\ SBN_O\ (\text{dBc/Hz}) + P_{TX} + [10 \times \text{LOG}_{10}\ (B)] + [10 \times \text{LOG}_{10}\ (Tx)] \\ &= -147 + 45 + [10 \times \text{LOG}_{10}\ (8 \times 10^3)] + [10 \times \text{LOG}_{10}\ (4)] \quad (12.18b) \\ &= -57\ \text{dBm/8 kHz} \end{aligned}$$

At this point, we need to jump ahead of ourselves briefly and consider that the base station receivers are also exposed to the remnants of the transmitted signals that pass through the receiver filter. These, in addition to the external high-level signals found during the RF survey, could also cause interference by means of receiver blocking or receiver IM. From Chapter 6, we know that when we have multiple interference mechanisms at play, we need to modify the interference threshold to take account of each contribution.

Interference per mechanism (dBm), $I'_T = I_T - [10 \times \text{LOG}_{10}\ (n)]$ (12.19a)

where

I_T = interference threshold, the total interference level at the base of the Rx antenna (dBm);

n = number of interference mechanisms simultaneously occurring.

In this case, we should have no remaining transmitter IM due to the design steps that we have taken to mitigate it, but we could have interference caused by transmitter sideband noise, receiver blocking, or receiver IM. So, we will assume the number of interference mechanisms present is 3 and modify the interference threshold accordingly.

$$\text{Interference per mechanism (dBm), } I'_T = I_T - [10 \times \text{LOG}_{10}\ (n)]$$
$$= -125.5 - [10 \times \text{LOG}_{10}\ (3)] \qquad (12.19b)$$
$$= -130.3 \text{ dBm}$$

We have now calculated the total transmitter sideband noise level on the transmit antenna at the nearest receive frequency (*Tx SBN*) and the maximum permissible level (I'_T) that we can tolerate at the base of the receive antenna for 1 dB of receiver desensitization when all transmitters are operating. So, the transmitter sideband noise suppression that we need between the base of the transmit antenna and base of the receive antenna is simply the difference between the two.

$$\text{Transmitter sideband noise suppression at receive frequency (dB)}$$
$$= Tx\ SBN - I'_T$$
$$= -57 - (-130.3) \qquad (12.20)$$
$$= 73.3 \text{ dB}$$

Of this, 60 dB is achieved simply by the isolation between the transmit and receive antennas, so only the remaining 13.3 dB is required from the cavity filters in each arm of the transmitter combiner.

To wrap up these calculations pictorially then, the cavity filter frequency response plot for the 152.525-MHz channel in our transmitter combiner would be as shown in Figure 12.15. Note that an extra 5 dB has been added to the calculated rejection figures to account for temperature drift and aging. This only considers the case between the highest transmit frequency and the lowest receive frequency. We could do the same calculation for the other three channels in the transmitter combiner but given their offsets to the nearest receive frequency are quite similar, the answers would be more or less the same.

Finally, presenting the transmitter sideband noise suppression calculation on a diagram like Figure 12.16 makes it easier to see what is going on at each point in the system.

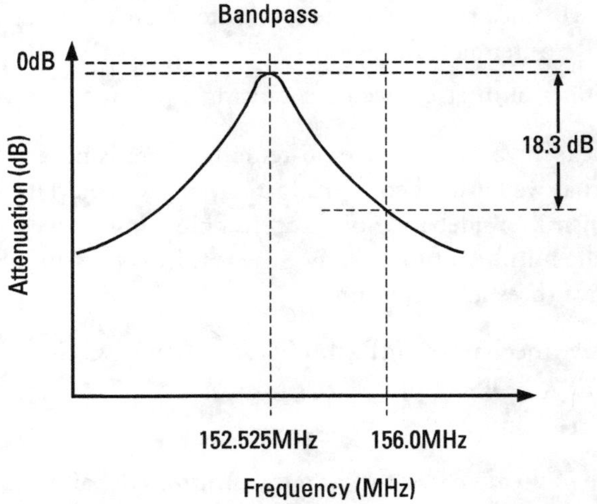

Figure 12.15 Transmitter combiner frequency response for a 152.525-MHz channel.

Table 12.8
Hilltop Radio Station Transmitter Combiner Specification

Parameter	**Design Limit**
Number of transmit channels	4
Filter characteristic	Bandpass
Transmitter frequencies	F_1 = 151.0 MHz F_2 = 151.375 MHz F_3 = 151.725 MHz F_4 = 152.525 MHz
Insertion (combining) loss	≤4.0 dB
Tx port return loss in Tx passband	Better than 22 dB
Isolation between combined transmitter outputs	80 dB
Single-carrier CW RF input power per channel	100W (+50 dBm)
Offset to closest receive frequency (lowest receive frequency to highest transmit frequency)	4.0 MHz
Sideband noise suppression at nearest receive frequency	18.3 dB
Input connectors (transmitter side)	Type N Female
Output connector (antenna side)	7/16DIN Female

Note the −118.7-dBm figure is the total noise and interference floor referred to the base of the receive antenna (N_X) that we worked out earlier during the interference threshold calculation.

The last step is to specify the transmitter combiner performance that we need, which is best done by means of Table 12.8. Note that 5 dB has been added to the filter rejection to account for temperature drift and aging.

To finish this part of the design, Figure 12.17 is a schematic drawing of our final transmitter combining system.

One thing to highlight quickly here is the schematic neatly shows the benefits of designing for 60-dB transmit to receive antenna isolation. With that, only another 18.3 dB of attenuation (including the 5-dB margin) is required

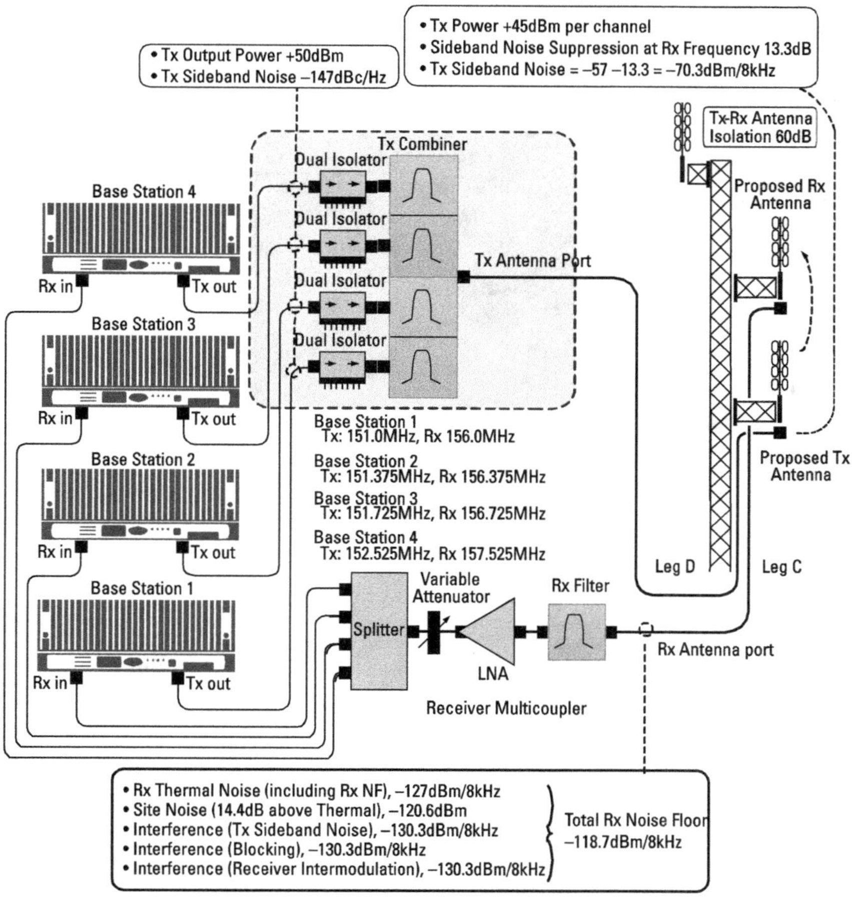

Figure 12.16 Transmitter combiner sideband noise suppression.

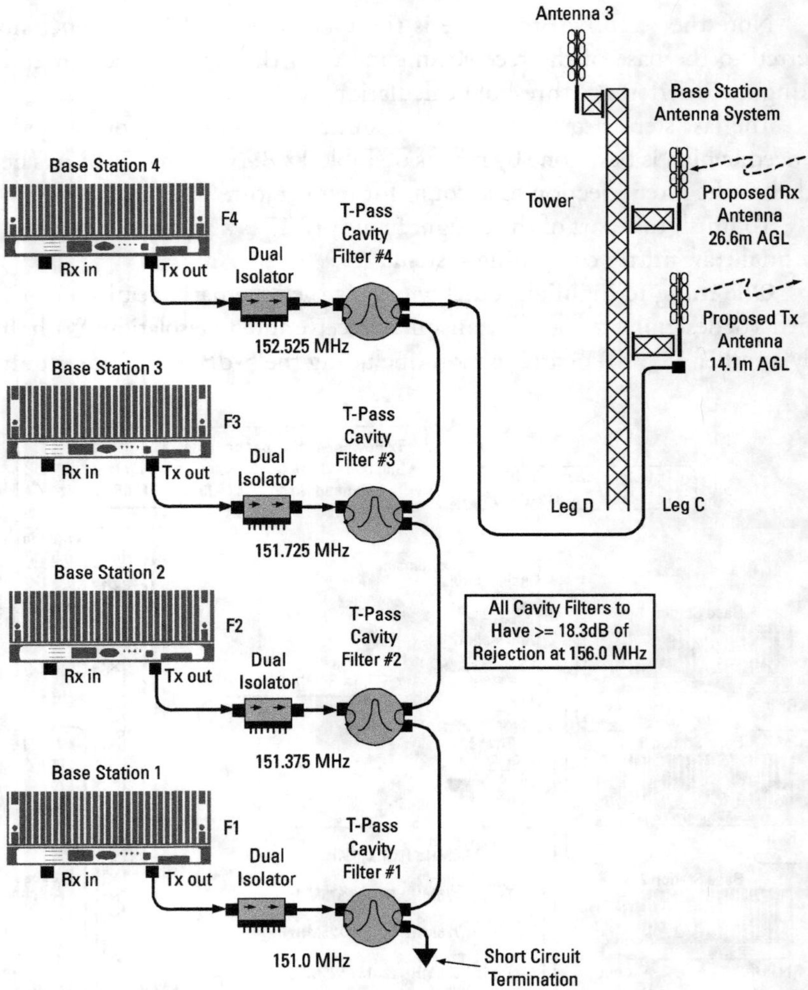

Figure 12.17 Hilltop Radio Station transmitter combining system.

in the cavity filters to suppress the total transmitter sideband noise below the interference threshold (–130.3 dBm) at the nearest receive frequency (156.0 MHz). This can be achieved with a single cavity filter in each arm of the transmitter combiner, thereby minimizing the losses, size, and cost of the system.

12.5 Receiver Multicoupler Design

Now that the positions of the antennas on the Hilltop Radio Station tower have been finalized, and the transmitter combining system has been designed, the next step is to specify the receiver multicoupler system according to the design rules given in Chapter 9 so that it allows all four receivers to be fed from the common receive antenna with no loss of sensitivity.

For the purposes of this design, we will define the receiver multicoupler as the receiver filter (also known as a preselector or roofing filter) and the following distribution system that amplifies and splits the received signal between the connected receivers as shown in Figure 12.18. It just makes our life easier because they need to be considered together from a design point of view.

In conjunction with good frequency planning and optimal antenna system design, the receiver multicoupler is our defense against local interference generated by all the transmitters at the site.

Let us now apply the design rules to design the receiver multicoupling system. Rule 1 emphasizes the importance of using separate transmit and receive antennas for multichannel repeater site multicoupling systems, as we have already done in the antenna system design. Next, applying rules 2 and 3 means that we can use a standard multicoupler because our transmit and receive frequencies are in distinct groups and there are no other transmitters at the site with frequencies that fall in between the highest and lowest receive

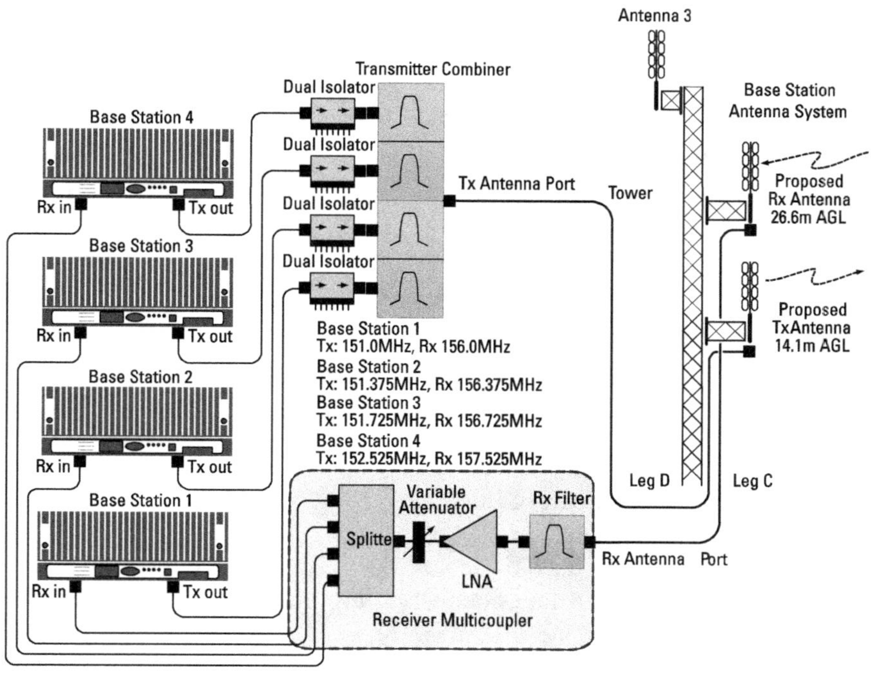

Figure 12.18 Four-channel receiver multicoupling system.

frequencies. Furthermore, as our system operates in the VHF band, there is no benefit in using a tower top amplifier (TTA) given the level of ambient site noise measured during the RF survey was 12.0–14.4 dB above thermal. Lastly, as the RF survey found no high-level transmit carriers within 150 kHz of the receive frequencies, so there is no need to use crystal filters in the receiver multicoupler design either.

12.5.1 Gain Setting

The general starting point is to set the gain of the multicoupler such that the sensitivity of the receivers is maintained. In other words, the sensitivity of each receiver connected to the multicoupler is the same as it would be if one of the receivers was connected directly to the base of the receive antenna. What that means is that the cascaded noise figure of the overall receiving system lineup (antenna cable, receiver filter, LNA, attenuator, splitter, and base-station receiver) needs to be the same as the 8-dB noise figure of the base station receiver alone.

The noise figure of the multicoupler is determined by the variable attenuator setting. So, let us set the variable attenuator to 11 dB as shown in Figure 12.19 and see what happens to the overall receiving system noise figure.

As previously covered in Chapter 9, the cascaded noise figure of the multicoupler lineup is calculated using the following formula.

$$\text{Cascade noise figure, } F = F_1 + [(F_2 - 1)/G_1] + [(F_3 - 1)/(G_1 \times G_2)] + [(F_4 - 1)/(G_1 \times G_2 \times G_3)]\ldots \quad (12.21a)$$

where (all quantities are pure ratios with no unit):

F_1 = noise figure of first part of the site receiving system (nearest the receive antenna);

G_1 = gain of first part of the site receiving system;

F_2 = noise figure of second part of the site receiving system;

G_2 = gain of second part of the site receiving system;

F_3 = noise figure of third part of the site receiving system;

G_3 = gain of third part of the site receiving system;

F_4 = noise figure of fourth part of the site receiving system.

Applying this to our four-channel Hilltop Radio Station site receiving system by using the gains, noise figures, and losses from Figure 12.19, we get:

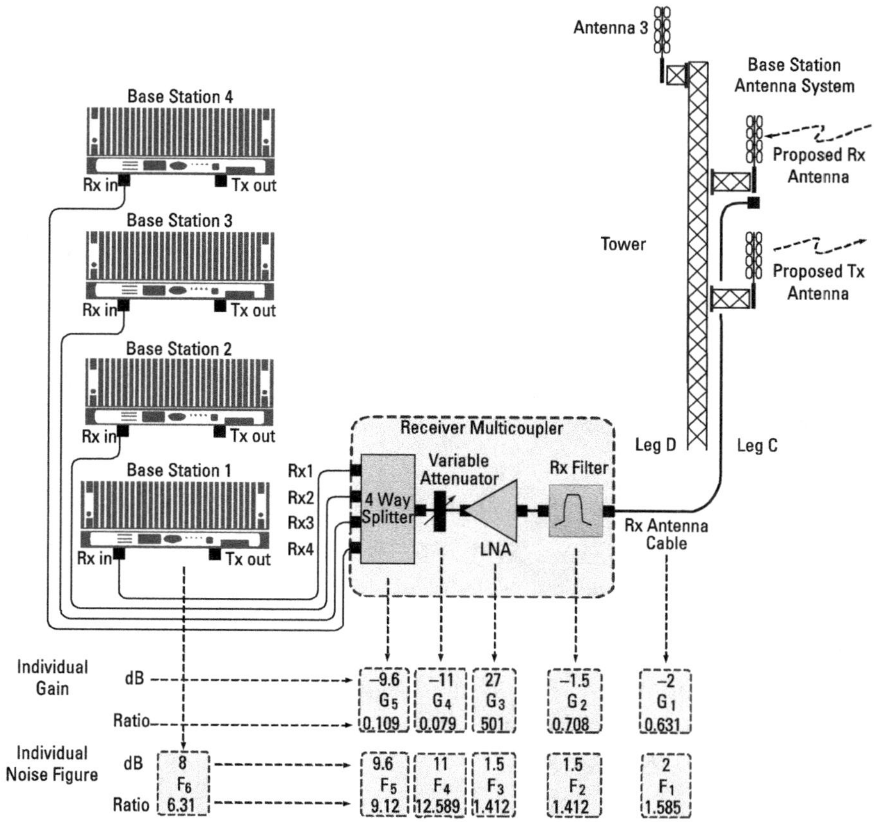

Figure 12.19 Receiver multicoupler setup to maintain the receiver noise figure.

Cascade noise figure, $F = F_1 + [(F_2 - 1)/G_1] + [(F_3 - 1)/(G_1 \times G_2)]$
$+ [(F_4 - 1)/(G_1 \times G_2 \times G_3)] + [(F_5 - 1)/(G_1 \times G_2 \times G_3 \times G_4)]$
$+ [(F_6 - 1)/(G_1 \times G_2 \times G_3 \times G_4 \times G_5)]$
$= 1.585 + [(1.412 - 1)/0.631] + [(1.412 - 1)/(0.631 \times 0.708)]$
$+ [(12.589 - 1)/(0.631 \times 0.708 \times 501)]$ (12.21b)
$+ [(9.12 - 1)/(0.631 \times 0.708 \times 501 \times 0.079)]$
$+ [(6.31 - 1)/(0.631 \times 0.708 \times 501 \times 0.079 \times 0.109)]$
$= 1.585 + 0.653 + 0.922 + 0.052 + 0.459 + 2.756$
$= 6.427$

Converting this cascaded noise figure to decibels, that is:

$$\text{Noise figure (dB)} = 10 \times \text{LOG}_{10}\ (F) \tag{12.22a}$$

where F = cascade noise figure as a ratio.

So

$$\begin{aligned}\text{Noise figure (dB)} &= 10 \times \text{LOG}_{10}\ (F)\\ &= 10 \times \text{LOG}_{10}\ (6.427)\\ &= 8.08 \text{ dB}\end{aligned} \tag{12.22b}$$

Hence, 11 dB is the correct attenuator setting to maintain the overall noise figure of the receiving system. If all the losses and gains through the system are added, this results in a gain through the multicoupler of 2.9 dB.

12.5.2 Calculating the Receiver Filter Requirements

The receiver filter prevents receiver blocking by rejecting high-level unwanted signals from colocated transmitters at the site and manages receiver IM by providing additional rejection of specific high-level unwanted signals that would otherwise mix in the receiving system and generate on-channel IM products.

In effect, the receiver filter is our defense against local interference that could otherwise deafen one or more of the connected receivers and limit their ability to hear weak signals from distant users in the field.

To start with, let us calculate the rejection required in the receiver filter to prevent any of the four receivers connected to our multicoupler being deafened because of receiver blocking. The relevant parameters we need to do this from some of our earlier calculations are:

- Transmitter power (per channel at the transmit antenna input), P_{TX} = +45 dBm. This is the transmit power per channel after the transmitter combiner loss (3 dB) and the RF cabling loss (2 dB);
- Transmit to receive antenna isolation = 60 dB;
- Receiver equivalent noise bandwidth, B = 8 kHz;
- Cascaded noise figure of the susceptible receiving system, $Rx\ NF$ = 8.08 dB;
- Total site noise floor, including site noise, $N_T = -119.7$ dBm;
- Total site noise and interference floor, $N_X = -118.7$ dBm;
- Interference threshold, $I_T = -125.5$ dBm;
- Rayleigh faded carrier-to-noise and interference ratio, $C_F\ /(N + I)$ = 15.2 dB;

- Static carrier-to-noise ratio, $(C_S / N) = 8$ dB;
- Dynamic receiver sensitivity = −103.5 dBm.

Generally, the blocking performance of a base station receiver is around 105 dB for DAQ3.0. However, this is a lab measurement made with constant level wanted and interfering signals. In practice, the wanted signal (a weak signal from a distant user in the field) will experience Rayleigh fading and vary in level, whereas the interfering signals will not as they are coming from colocated transmit antennas at the site. To account for this, we must reduce the blocking performance by an amount equal to the difference between the static carrier to noise ratio (C_S / N) and the dynamic carrier-to-noise and interference ratio $(C_F / (N + I))$. This is 7.2 dB, meaning that the blocking performance becomes 97.8 dB. Also, we have four transmitters in this multichannel system, plus the colocated 136.0-MHz transmitter currently operating on the same tower (see the tower mapping performed earlier), and the 146.0-MHz transmitter picked up during the RF survey that are all present at the receive antenna at similar levels (note that the 161.875-MHz and 166.225-MHz interfering signals are present at much lower levels and so are not the main receiver blocking threat). Each of these high-level interfering signals contributes to the problem, so we need to account for them by reducing the blocking performance by a further 9 dB (see design rule 6) to 88.8 dB. In addition, the base station receivers are also subject to interference from transmitter sideband noise and (let us assume) receiver IM too. That means there are three interference mechanisms present for which we account by reducing the blocking performance by another 4.8 dB, to 84.0 dB. What that means is that the maximum interfering signal level that we can tolerate at any base station receiver connected to the multicoupler is 84.0 dB above the dynamic receiver sensitivity.

$$\begin{aligned}
&\text{Maximum interfering signal at receive antenna (dBm)}\\
&= \text{Receiver sensitivity}\\
&+ \text{Blocking} \qquad (12.23)\\
&= -103.5 + 84\\
&= -19.5 \text{ dBm}
\end{aligned}$$

Next, we need to know the level of the interfering transmit signal that we need to reject. In this case, it makes sense to just consider the transmit signals from our associated transmitter combiner, as these are the highest level interfering signals arriving at the receive antenna and the closest to the receive frequencies. Now, as we know the transmitter power per channel is +45 dBm and the transmit to receive antenna isolation is 60 dB, the level of each transmit signal at the receive antenna output would be −15 dBm. So, as we have just calculated that the maximum interfering signal level that we can tolerate at the receive antenna is −19.5 dBm, the rejection required in the receive filter at the closest transmit frequency is simply the difference between the two.

Receive filter rejection at closest transmit frequency (dB)
= Interfering signal level at receive antenna (dBm)
− Maximum interfering signal level at receive antenna (dBm) (12.24)
= −15 − (−19.5)
= 4.5 dB

This relatively easy requirement again emphasizes the benefit of maximizing the transmit to receive antenna isolation, which is effectively providing most of the blocking rejection for free.

What is going on is much easier to visualize on a diagram like Figure 12.20, as usual.

This example is just for associated transmitter signals. The blocking rejection required in the receiver filter for unrelated high-level signals from other colocated transmitters at the site can be calculated in the same way; all you have to know is that their level at the receive antenna recorded during the RF survey. If we did that, the rejection we would require in the receiver filter to prevent blocking due to any of the interfering transmit signals present at the receive antenna would be as given in Table 12.9. Note again that an extra 5 dB has been added to all the calculated rejection figures to account for filter temperature drift and aging.

Having calculated the receiver filter rejection necessary to prevent blocking, we next need to carry out receiver IM analysis and see whether any additional attenuation of specific interfering signals is required too.

Receiver IM differs from transmitter IM in that all interfering signals present at the receive antenna must now be considered. So, if we include the transmit and receive frequencies of our Hilltop Radio Station system, plus the in-band interfering signals found during the tower mapping and RF survey, we end up with the data shown in Table 12.10.

What Table 12.10 means is we are searching for IM products generated by combinations of these transmit frequencies that fall within the bandwidth of any of the receivers.

For systems like ours where we have 60 dB of isolation between the transmit and receive antennas, receiver multicoupler design rule 7 tells us that we only need to search for third, fifth, and seventh-order receiver IM products within a search bandwidth of:

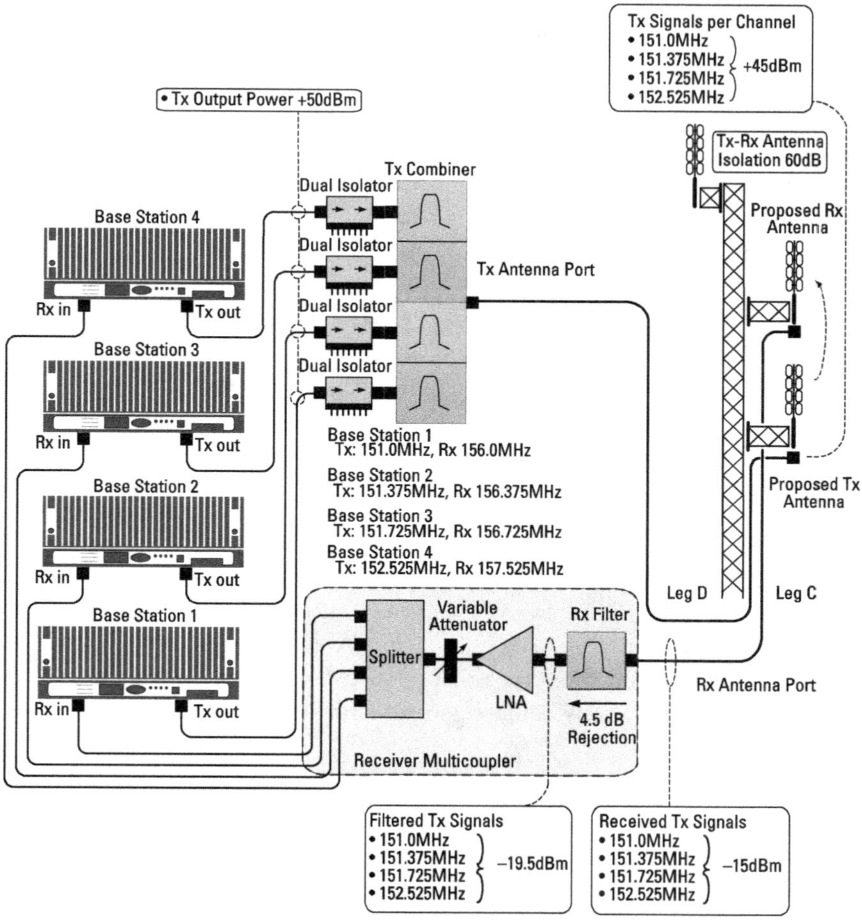

Figure 12.20 Hilltop Radio Station receiver filter blocking rejection.

$$\text{Search bandwidth (kHz)} = (n + 1) \times \text{Channel bandwidth} \tag{12.25}$$

where

n = maximum order of IM to be searched for;

Channel bandwidth = 12.5 kHz (typical for LMR systems).

So, for a maximum of seventh-order IM and a 12.5-kHz channel bandwidth, the search bandwidth is 100 kHz (+/−50 kHz around each receive frequency) because the occupied bandwidth of an IM product spreads according to its order.

Table 12.9
Receiver Filter Blocking Requirements

Interfering Transmitter Frequency	Receiver Filter Blocking Rejection
151.0 MHz	9.5 dB
151.375 MHz	9.5 dB
151.725 MHz	9.5 dB
152.525 MHz	9.5 dB
136.0 MHz	16.5 dB
146.0 MHz	16.5 dB
161.875 MHz	2.5 dB
166.225 MHz	0 dB

Running the receiver IM analysis shows that we have direct third, fifth, and seventh-order IM products falling on all the receive frequencies as summarized in Table 12.11.

Clearly, something needs to be done to prevent these IM products from deafening the receivers, so let us look at a couple of different mitigation strategies that could be used.

First, let us think about the third-order IM products affecting the 156.0-MHz receive frequency, which are generated via the mechanisms:

$$(2 \times 146.0 \text{ MHz}) - (136.0 \text{ MHz}) = 156.0 \text{ MHz} \tag{12.26}$$

$$(2 \times 151.0 \text{ MHz}) - (146.0 \text{ MHz}) = 156.0 \text{ MHz} \tag{12.27}$$

Table 12.10
Frequencies for Hilltop Radio Station Receiver IM Analysis

Transmit Frequencies		Receive Frequencies
Transmit signals from our system	151.0 MHz	156.0 MHz
	151.375 MHz	156.375 MHz
	151.725 MHz	156.725 MHz
	152.525 MHz	157.525 MHz
Other in-band transmitters on same tower	136.0 MHz	
Other in-band transmitters from other towers (picked up during the RF survey)	146.0 MHz	
	161.875 MHz	
	166.225 MHz	

Table 12.11
Hilltop Radio Station Receiver IM Results

IM Order	Number of Receiver IM Products at Each Receive Frequency							
	156.0 MHz		156.375 MHz		156.725 MHz		157.525 MHz	
	Direct	Indirect	Direct	Indirect	Direct	Indirect	Direct	Indirect
Third	2	0	0	1	1	1	2	0
Fifth	1	2	2	4	2	0	1	2
Seventh	5	13	5	9	5	14	4	7

Taking the first mechanism, this is generated by the 136.0-MHz carrier from antenna 3 on the same tower, plus the strong 146.0-MHz carrier from a neighboring tower picked up during the RF survey, as shown in Figures 12.21 and 12.22.

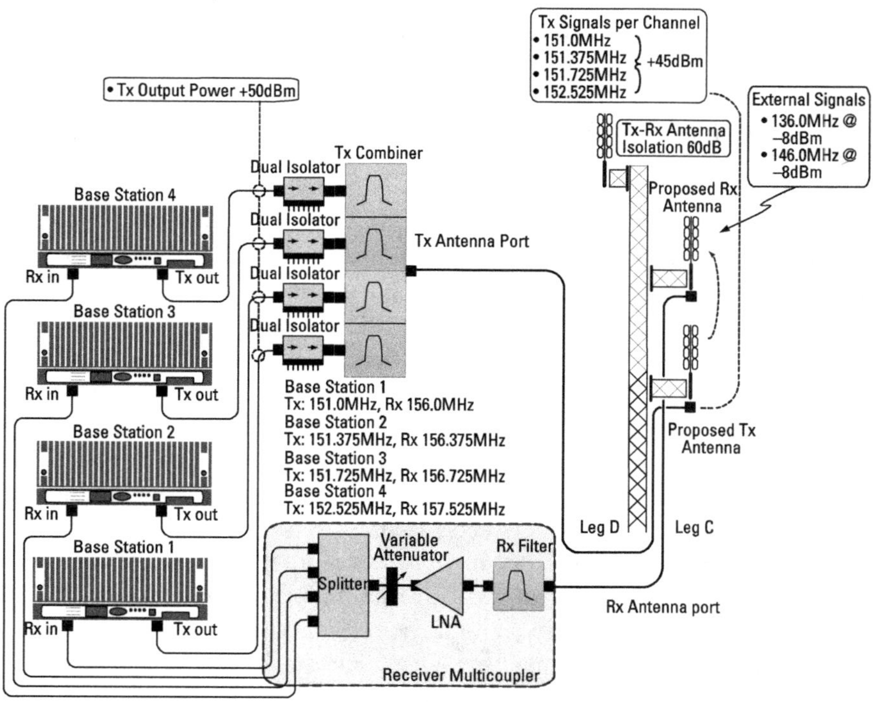

Figure 12.21 Third-order IM caused by 136.0-MHz and 146.0-MHz HLCs.

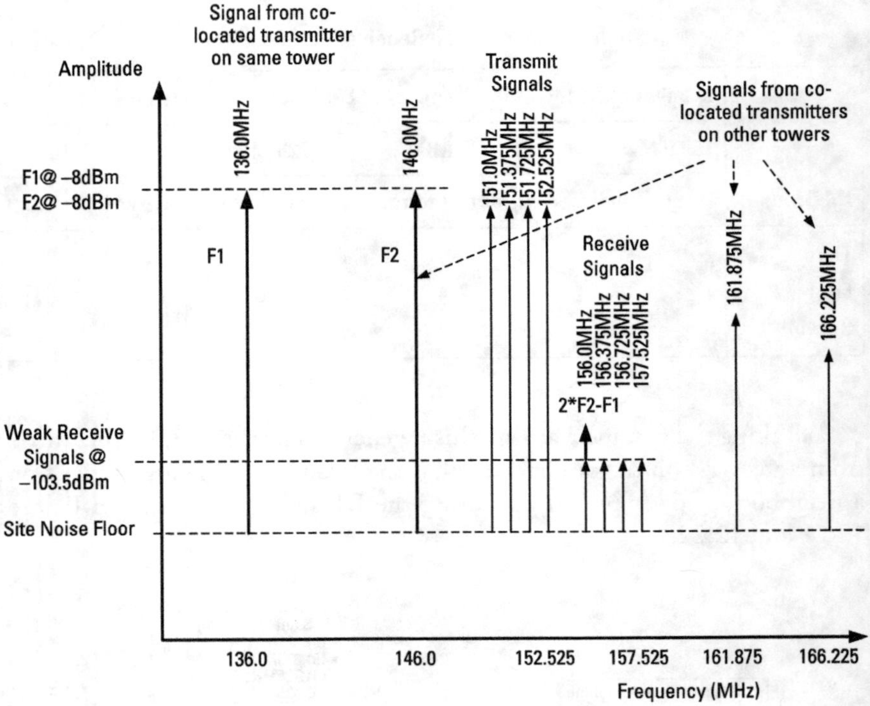

Figure 12.22 Third-order IM at 156 MHz.

These interfering signals are 10 MHz or more away from the receive band, so are likely to be fully rejected by the bandpass receiver filter. However, because of their extremely high levels, they could generate PIM on the receive antenna. In cases like this where the potential third-order IM is caused by two or more very high-level external signals with high occupancy, a frequency plan change should be seriously considered. For example, if we simply moved the affected 156.0-MHz receive frequency by 100 kHz, the third-order product would no longer fall in band. Another advantage of this, if we look back to the third-order IM mechanisms present at 156.0 MHz, is that it would mitigate the (2 × 151.0 MHz – 146.0 MHz) mechanism too. So, having made sure that we are not causing any cochannel or adjacent channel interference with neighboring sites, let us go ahead with this frequency change. Table 12.12 shows our updated frequency plan for the Hilltop site.

Having made a frequency change now means that we have to rerun the IM analysis using the frequencies in Table 12.13.

Table 12.12
Revised Frequencies for Hilltop Radio Station

Channel	Frequency Pair	Transmit Frequency	Receive Frequency
1	F_5	151.1 MHz	156.1 MHz
2	F_{12}	151.375 MHz	156.375 MHz
3	F_{17}	151.725 MHz	156.725 MHz
4	F_{21}	152.0 MHz	157.0 MHz

With these frequencies, we get the following receiver IM results in Table 12.14.

Comparing these results with those for the initial frequency plan (see Table 12.12) shows a good improvement, with fewer third and fifth-order IM products generated at all the receive frequencies.

The effect of this frequency change on the previous blocking calculations is insignificant as the blocking rejection required at 156.1 MHz will be the virtually the same as at the previous frequency of 156.0 MHz. Other than changing the center frequency of the cavity filter, there is no change necessary to the transmitter combiner design either because the sideband noise rejection needed at 151.1 MHz will be similar to that required at 151.0 MHz.

Table 12.13
Revised Frequencies for Hilltop Radio Station Receiver IM Analysis

Transmit Frequencies		Receive Frequencies
Transmit signals from our system	151.1 MHz	156.1 MHz
	151.375 MHz	156.375 MHz
	151.725 MHz	156.725 MHz
	152.525 MHz	157.525 MHz
Other in-band transmitters on same tower	136.0 MHz	
Other in-band transmitters from other towers (picked up during the RF survey)	146.0 MHz	
	161.875 MHz	
	166.225 MHz	

Table 12.14
Revised Hilltop Radio Station Receiver IM Results

IM Order	Number of Receiver IM Products at Each Receive Frequency							
	156.1 MHz		156.375 MHz		156.725 MHz		157.525 MHz	
	Direct	Indirect	Direct	Indirect	Direct	Indirect	Direct	Indirect
Third	0	2	0	0	0	2	1	0
Fifth	1	3	1	6	1	3	0	5
Seventh	4	12	6	9	5	13	4	10

Just for completeness, Figure 12.23 shows the RF environment diagram for the Hilltop Radio Station site with the new transmit and receive frequency pair.

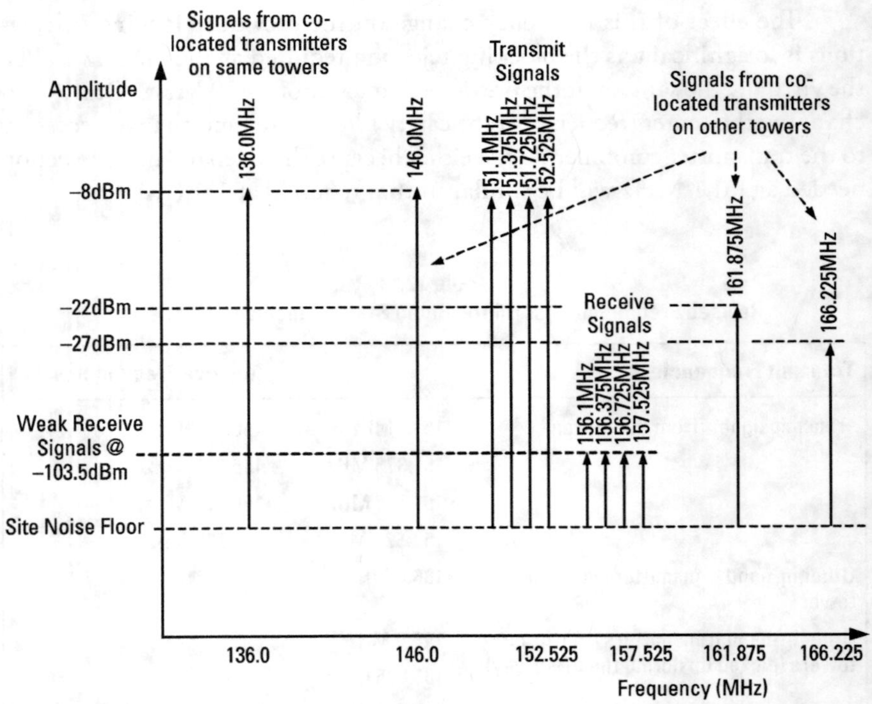

Figure 12.23 Hilltop Radio Site RF environment after frequency change.

Now, let us think about the third-order IM direct hit affecting the 157.525-MHz receive frequency, which is shown in Figures 12.24 and 12.25, and generated via the mechanism:

$$(2 \times 161.875 \text{ MHz}) - (166.225 \text{ MHz}) = 157.525 \text{ MHz} \tag{12.28}$$

Compared to the 136.0-MHz and 146.0-MHz HLCs, the 161.875-MHz and 166.225-MHz interfering signals are much closer to the receive band and so may not be fully rejected by the receive filter. From our earlier calculations, we know the receiver filter only has to offer minimal suppression of the 161.875-MHz and 166.225-MHz external interfering signals to prevent them causing receiver blocking (see Table 12.10). However, we now need to see if this is enough to prevent them generating IM at a level above the interference threshold.

To calculate the level of the IM product generated by the 161.875-MHz and 166.225-MHz external signals, we first need to work out the cascaded

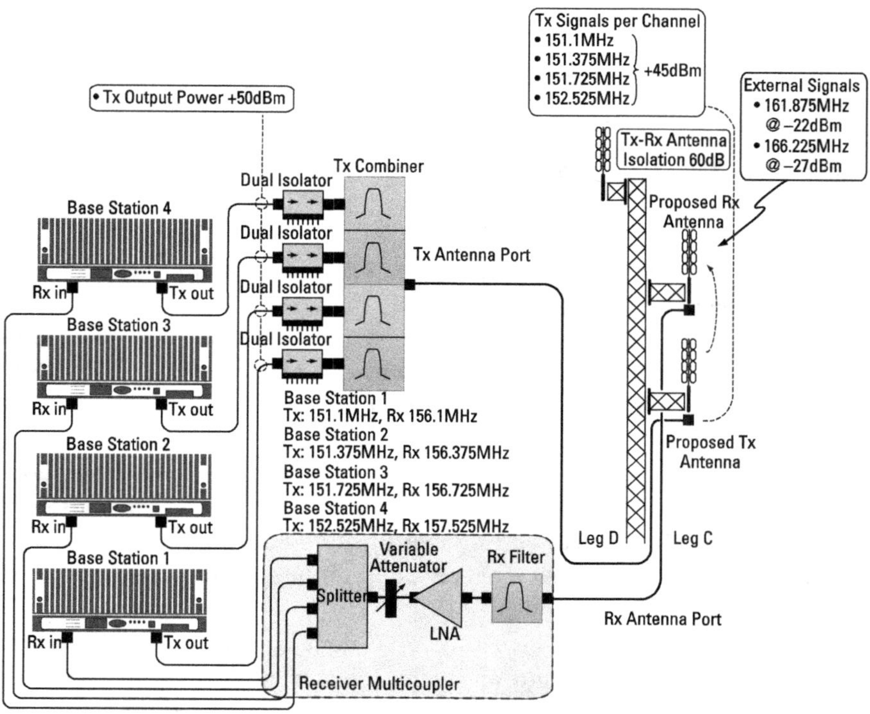

Figure 12.24 Third-order IM caused by 161.875-MHz and 166.225-MHz HLCs.

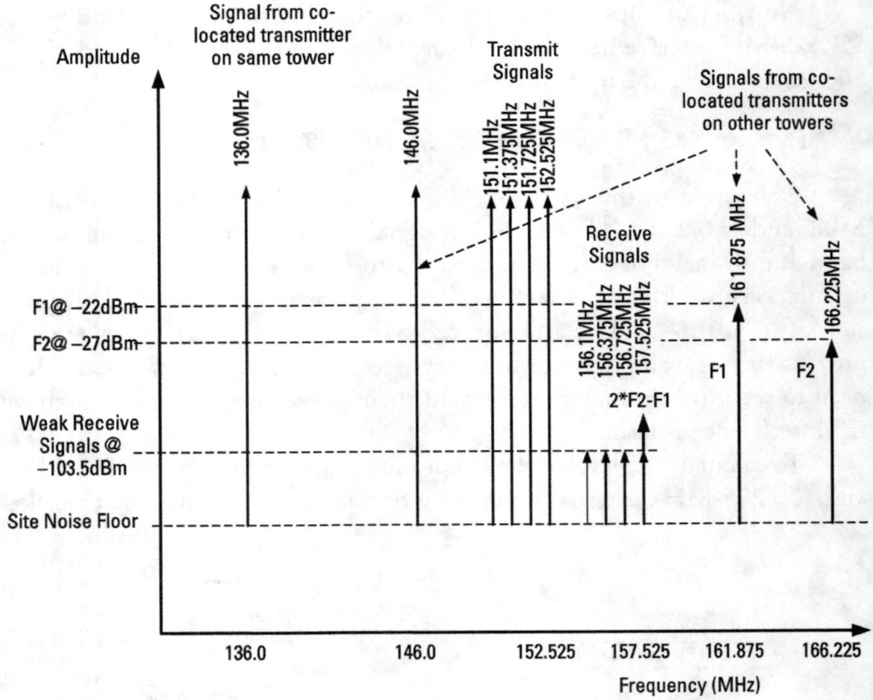

Figure 12.25 Third-order IM at 157.525 MHz.

intercept point of our receiver multicoupler set up to maintain the 8-dB noise figure as shown in Figure 12.26.

The only two parts of the receiving system that have their own third-order intercept points are the base station receivers themselves and the LNA. The base station receivers have an IM specification of 85 dB when two input signals of −107 dBm (1 mV) are applied, which translated means that the IIP3 of the base station receivers is +16 dBm. Additionally, the LNA specifications tell us that the OIP3 of the LNA is +40 dBm. This and the gains/losses of the other stages in the receiving system lineup are all we need to know to calculate the overall cascaded intercept point of the system using the method shown in Figure 12.27.

The calculation from Figure 12.27 is done by starting from the base station receiver and working towards the antenna. First, take the +16-dBm IIP3 of the base station receiver and move forward towards the antenna one stage to work out the third-order intercept point at the input to the splitter by subtracting the splitter loss of 9.6 dB.

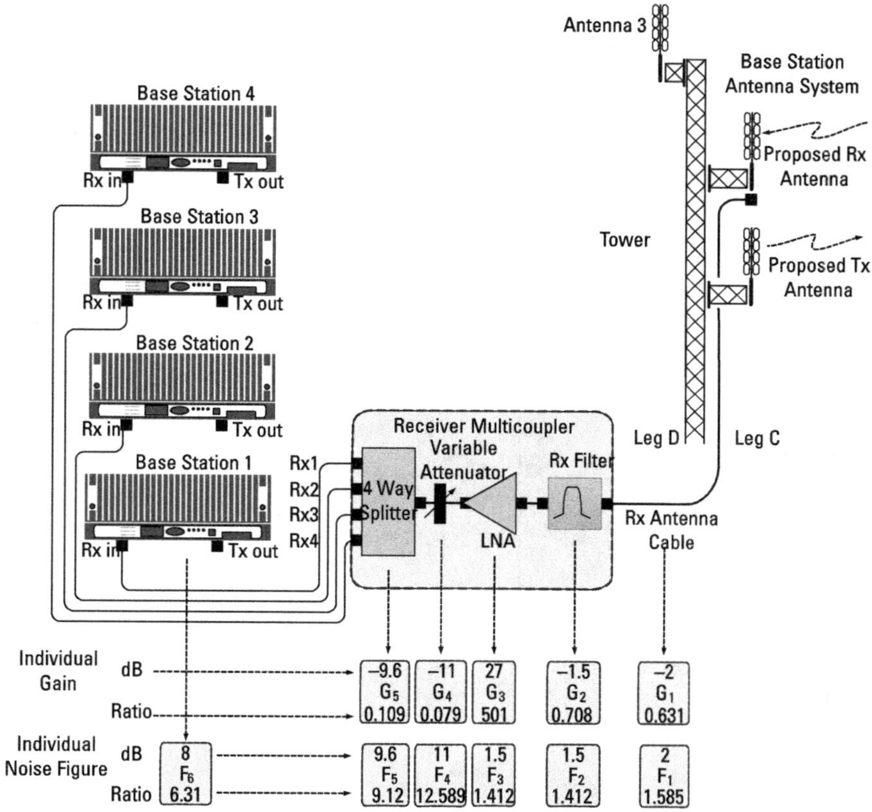

Figure 12.26 Receiver multicoupler setup to maintain the overall system noise figure.

$$\text{IIP3}_S\ (\text{dBm}) = \text{IIP3}_{\text{Rx}} - G_5 \qquad (12.29\text{a})$$

where

IIP3_S = input third-order intercept point at the splitter input (dBm);
IIP3_{Rx} = input third-order intercept point at the receiver input (dBm);
G_5 = splitter gain (dB).

So:

$$\begin{aligned}\text{IIP3}_S\ (\text{dBm}) &= \text{IIP3}_{\text{Rx}} - G_5 \\ &= +16 - (-9.6) \\ &= +25.6\ \text{dBm}\end{aligned} \qquad (12.29\text{b})$$

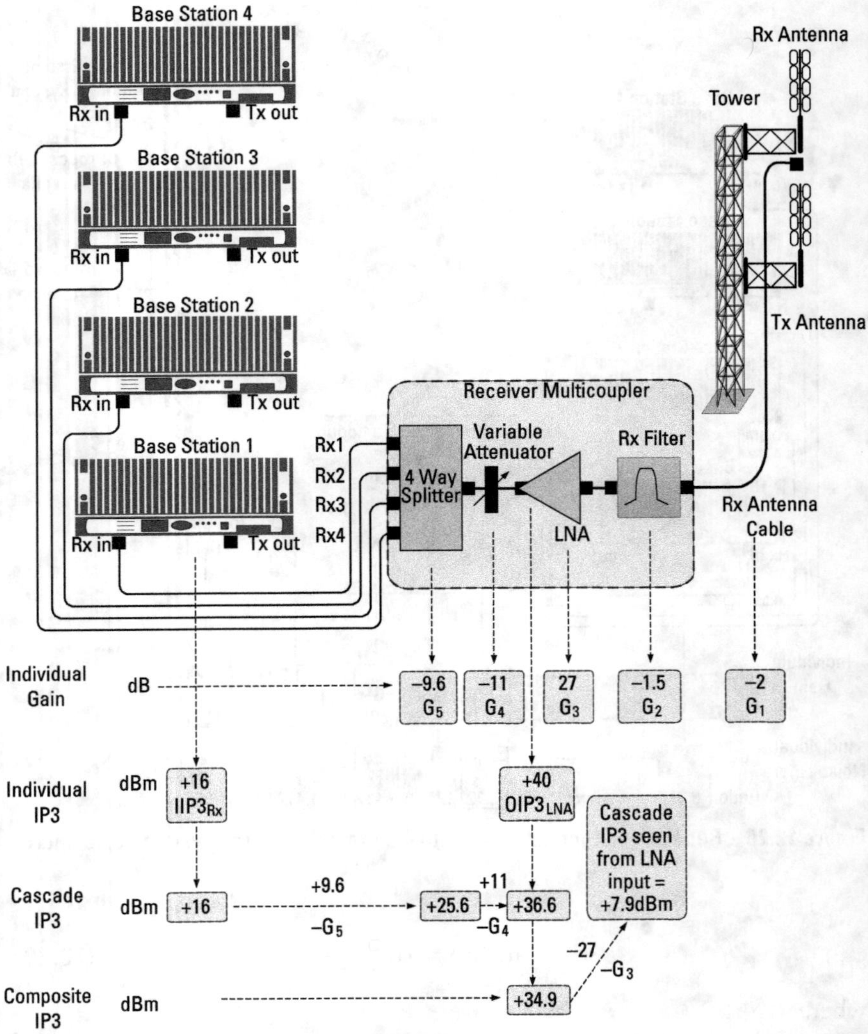

Figure 12.27 Cascaded IP3 of receiving system.

Next, we move another stage towards the antenna and calculate the IIP3 referred to the input of the 11-dB attenuator.

$$\text{IIP3}_{\text{ATT}}\ (\text{dBm}) = \text{IIP3}_S - G_4 \tag{12.30a}$$

where

IIP3_{ATT} = input third-order intercept point at attenuator input (dBm);

$IIP3_S$ = input third-order intercept point at splitter input (dBm);
G_4 = attenuator gain (dB).

So

$$\begin{aligned} IIP3_{ATT}\ (\text{dBm}) &= IIP3_S - G_4 \\ &= +25.6 + (-11) \\ &= +36.6\ \text{dBm} \end{aligned} \tag{12.30b}$$

At this point in the lineup, the overall third-order intercept point is also influenced by the output third-order intercept point (OIP3) of the LNAs shown in Figure 12.27. This is addressed by calculating the composite IP3 of the two as follows.

$$1/\text{Composite IP3} = 1/IIP3_{ATT} + 1/OIP3_{LNA} \tag{12.31a}$$

where

Composite IP3 = input third-order intercept point at splitter input (mW);
$IIP3_{ATT}$ = input third-order intercept point at attenuator input (mW);
$OIP3_{LNA}$ = input third-order intercept point at attenuator input (mW).

So

$$\begin{aligned} IIP3_{ATT}\ (\text{mW}) &= 10^{(IIP3_{ATT}\ (\text{dBm})/10)} \\ &= 10^{(+36.6/10)} \\ &= 4{,}571\ \text{mW} \end{aligned} \tag{12.31b}$$

and

$$\begin{aligned} OIP3_{LNA}\ (\text{mW}) &= 10^{(OIP3_{LNA}\ (\text{dBm})/10)} \\ &= 10^{(+40/10)} \\ &= 10{,}000\ \text{mW} \end{aligned} \tag{12.31c}$$

$$\begin{aligned} 1/\text{Composite IP3} &= (1/IIP3_{ATT}) + (1/OIP3_{LNA}) \\ &= (1/4{,}571) + (1/10{,}000) \\ &= 3.19 \times 10^{-4} \end{aligned} \tag{12.31d}$$

$$\begin{aligned} \text{Composite IP3} &= 10 \times \text{LOG}_{10}\ [1/(3.19 \times 10^{-4})] \\ &= +34.9\ \text{dBm} \end{aligned} \tag{12.31e}$$

and, finally, from this we subtract the gain of the LNA.

$$\text{IIP3}_{\text{LNA}}\ (\text{dBm}) = \text{Composite IP3} - G_3 \tag{12.32a}$$

where

IIP3_{LNA} = input third-order intercept point at LNA input (dBm);
Composite IP3 = third-order intercept point at LNA output (dBm);
G_3 = LNA gain (dB).

So

$$\begin{aligned}\text{IIP3}_{\text{LNA}}\ (\text{dBm}) &= \text{Composite IP3} - G_3 \\ &= +34.9 - (27) \\ &= +7.9\ \text{dBm}\end{aligned} \tag{12.32b}$$

Using this cascaded IP3 figure, we can now estimate the level of the third-order IM product generated in the receiving system by the 161.875-MHz and 166.225-MHz external interfering signals.

These signals were received at levels of −22 dBm and −27 dBm, respectively, and the loss they experienced through the receive antenna cable and receiver filter is 3.5 dB (see Figure 12.27). So, their levels at the same point in the lineup reduce to −25.5 dBm and −30.5 dBm.

To calculate the level of the $2 \times F_1 - F_2$ product that lands on the 157.525-MHz receive frequency, first notice the level of the 161.875-MHz interfering signal F_1 (−25.5 dBm) is 33.4 dB below the IP3 at the output of the receiver filter (+7.9 dBm). So, because the order of F_1 involved in this IM product is the second harmonic, the level of the $2 \times F_1 - F_2$ product drops by twice 33.4 dB below the IP3, to −58.9 dBm. In addition, the level of external interfering signal F_2 (−30.5 dBm) is 38.4 dB below the IP3 at the output of the receiver filter (+7.9 dBm), so because the order of F_2 involved in this IM product is the fundamental, the $2 \times F_1 - F_2$ product level drops by another 38.4 dB below the IP3 at the output of the receiver filter to −97.3 dBm. This is way above our desired dynamic sensitivity of −103.5 dBm and our interference threshold of −130.3 dBm as highlighted by Figure 12.28. In real terms, it means that this IM product is causing strong interference and severely limiting the distance from which a user can communicate with the site on the base station 4 receive frequency of 157.525 MHz.

The task now is to calculate the additional rejection needed in the receiver filter to suppress the $2 \times F_1 - F_2$ IM product down to the interference per mechanism threshold of −130.3 dBm.

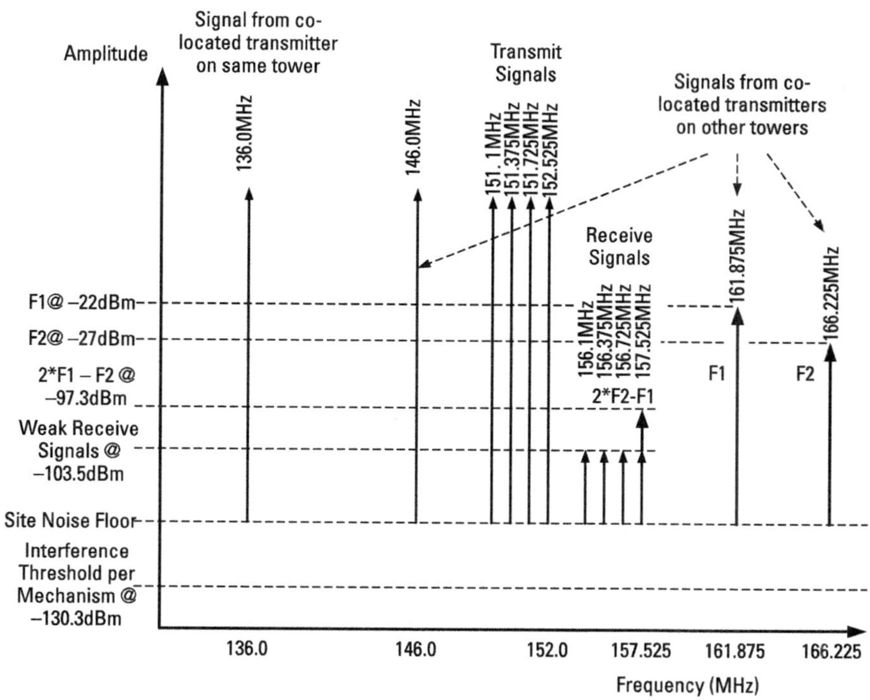

Figure 12.28 Third-order receiver IM level at 157.525 MHz.

Let us start by considering the rejection at 161.875 MHz, as that is the interfering signal closest to the receiver filter passband. If we were to design the receiver filter to provide 8 dB of rejection at 161.875 MHz, that would suppress the level of the $2 \times F_1 - F_2$ IM product by twice that amount (because the second harmonic of 161.875 MHz is involved) down to −113.3 dBm. If we design the receiver filter to also provide 17 dB of rejection at 166.225 MHz, then the level of the IM product reduces further by the same amount (because the fundamental of 166.225 MHz is involved) to the interference per mechanism threshold (I_T) of −130.3 dBm.

To wrap up these calculations pictorially in Figure 12.29, then the minimum receiver filter frequency response that we require would look like this. Note that an extra 5 dB has been added to the calculated rejection figures to account for temperature drift and aging.

Now, as we have throughout this example, let us show the receiver filter IM rejection calculation on a diagram (Figure 12.30) to see what is going on at each point in the system.

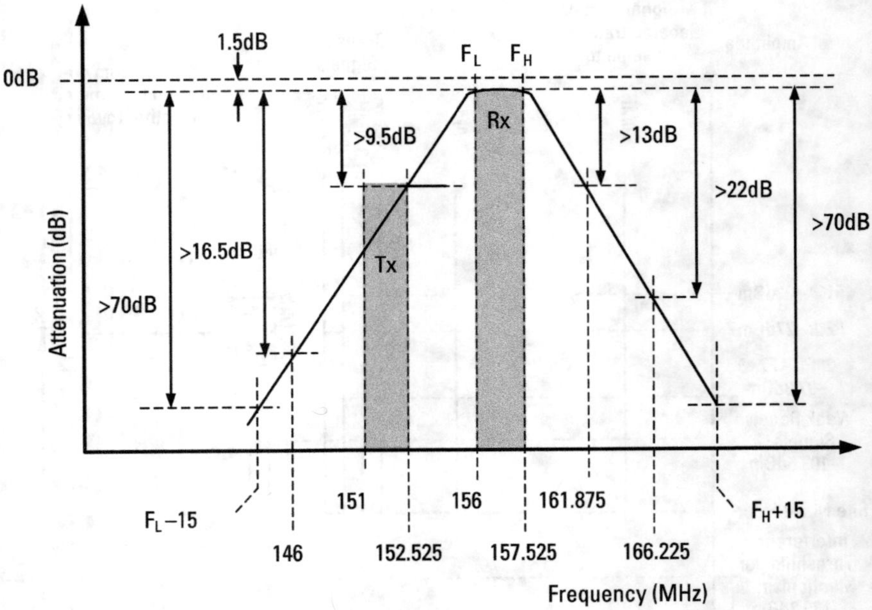

Figure 12.29 Receiver filter frequency response.

Finally, to complete our receiver multicoupler design, let us summarize the performance that we need from each part of the system in Table 12.15.

To finish this section, Figure 12.31 shows the schematic drawing of our final receiver multicoupling system.

12.6 Antenna System Monitor

As described in Chapter 10, antenna system monitors allow us to remotely keep an eye on the health of the transmitters and transmit antennas at the sites. Faults like a drop in transmitter power or a rise in antenna VSWR are highlighted in real time allowing proactive replacement of ailing parts before complete failure and loss of coverage occurs.

In the case of the Hilltop Radio Station system that we are designing here, a full channelized antenna system monitor is more appropriate as we are combining four transmitters on to the common antenna. A simpler composite power monitor would only provide measurement of total RF power

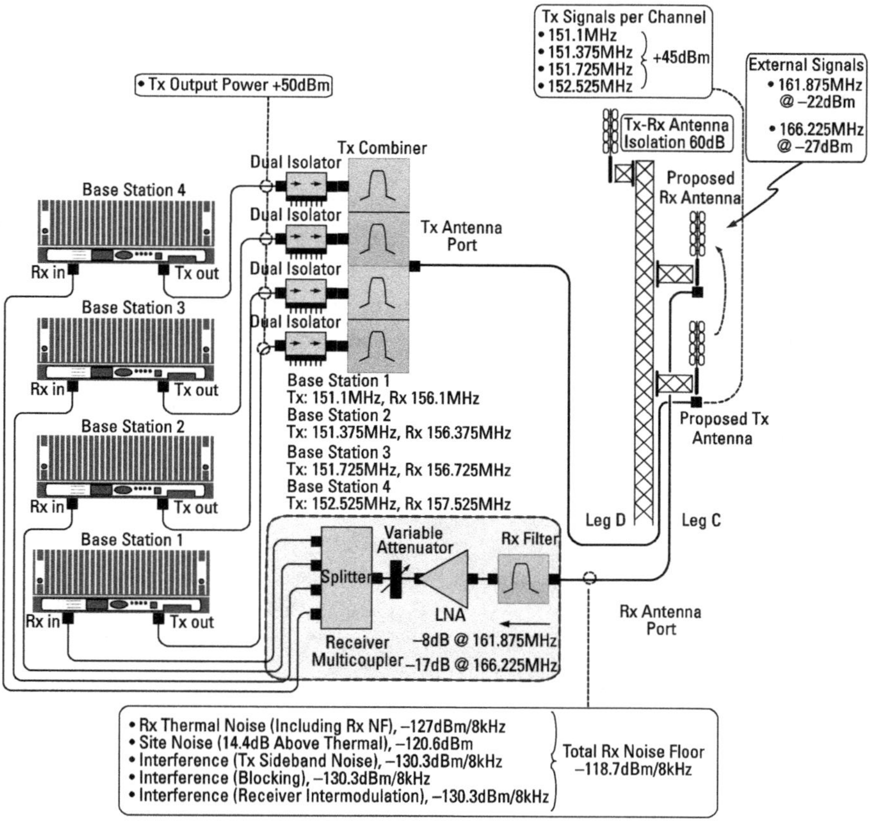

Figure 12.30 Receiver multicoupler third-order IM rejection.

and therefore be somewhat ambiguous given any number of the combined transmitters could be active at any one time.

Figure 12.32 shows where the channelized antenna system monitor would fit into our system.

The monitoring system uses a 30-dB directional coupler connected at the transmitter combiner output. The forward and reverse ports of the coupler are connected to the channelized antenna system monitor so the transmit power per channel and the transmit antenna VSWR can be calculated. The internet connection then allows remote monitoring of the measurements, and an alarm is activated if the reverse power or VSWR rises above a predetermined programmed level.

Table 12.15
Hilltop Receiver Multicoupler Specifications

Receiver filter	
Parameter	**Performance**
Filter characteristic	Bandpass
Passband	156.1 MHz (F_L) – 157.525 MHz (F_H)
In-band insertion loss	<1.5 dB
In-band input return loss	Better than 22 dB
Blocking rejection	>9.5 dB from 151.1 to 152.525 MHz
	>16.5 dB at 136.0 MHz
	>16.5 dB at 146.0 MHz
IM rejection	>13 dB at 161.875 MHz
	>22 dB at 166.225 MHz
Stopband	$<F_L - 15$ MHz, $> F_H + 15$ MHz
Minimum rejection in stopband	>70 dB
Input connector (antenna side)	Type N Female
Output connector (receiver side)	Type N Female
LNA	
Parameter	**Performance**
Gain	27 +/–1 dB
Noise figure	≤1.5 dB
Output IP3	≥+40 dBm
Input connector (receiver filter side)	Type N Female
Variable attenuator	
Parameter	**Performance**
Attenuation range	0 to 15 dB
Attenuation step	1 dB
Attenuator setting	11 dB
Splitter	
Parameter	**Performance**
Number of output ports	4
Splitter loss	9.6 +/– 1 dB
Output connector (receiver filter side)	BNC Female

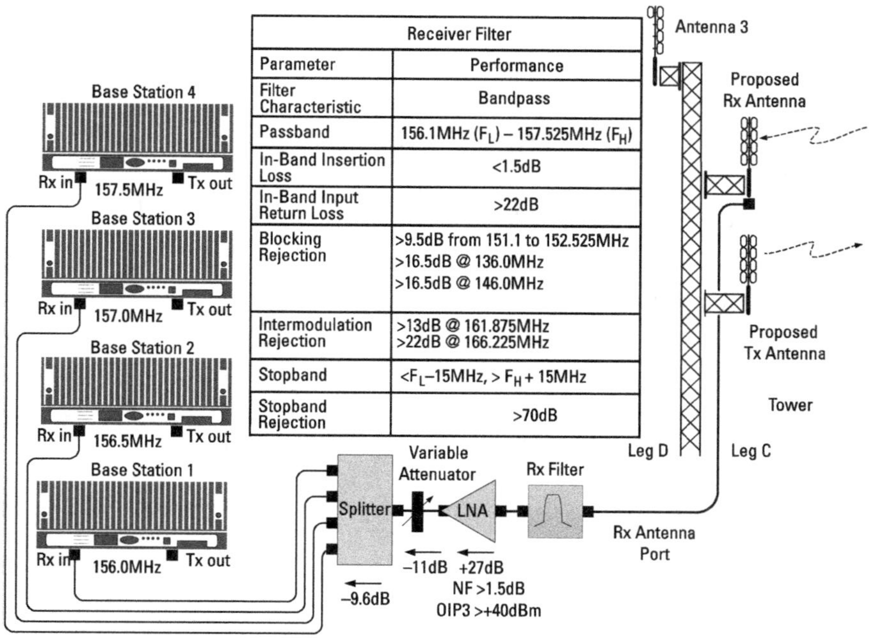

Figure 12.31 Hilltop Radio Station receiver multicoupling system.

12.7 Transmission Line System

Now that the antenna system, transmitter combiner, and receiver multicoupler have been designed, we need to connect these subsystems together. In this section, we will choose the appropriate types of transmission lines and connectors to use and also the related items such as surge suppressors and grounding devices.

Let us start by revising Figure 11.1 from Chapter 11, which shows the RF cables needed to connect the Hilltop Radio Station multichannel RF system.

To work out the length of the major cables, we need to look at the site plan drawn up during the site survey (see Figure 12.2) and the antenna system design derived earlier.

From these, we can see the length of the cable tray from the rack to the surge suppressors at the entry and exit plate will be 16m. In addition, the cable tray sits near the ceiling level, 2m above the rack. Adding these together, the Tx and Rx surge suppressor jumpers both need to be 18m long.

Next, we need to calculate the lengths of the Tx and Rx antenna lines. The proposed transmit and receive antenna base heights from our antenna

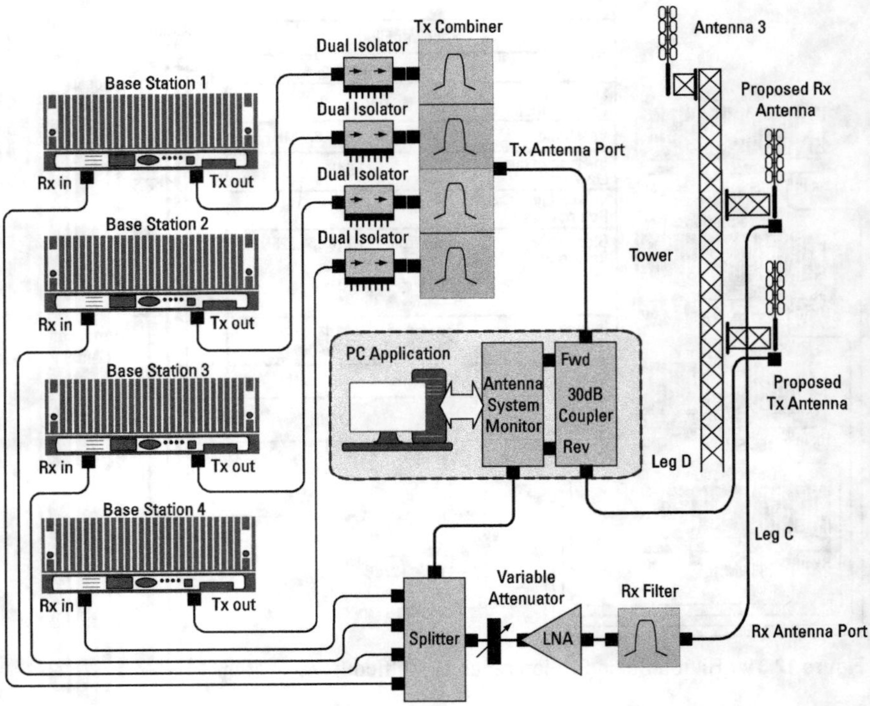

Figure 12.32 Hilltop Radio Station channelized antenna system monitor.

system design in Figure 12.12 are 14.1m and 26.6m, respectively. In addition, the site plan (Figure 12.2) told us the length of the ice bridge from the equipment shelter to the tower is 12m. Finally, from the site survey, the distance from where the ice bridge joins the tower to leg C where the antennas are mounted is 1.5m. At the antenna end, though, we have the transmit and receive antenna jumpers, which are 3-m-long lengths of a thinner 1/2-inch cable designed to relieve the stress of connecting to the antennas. The antennas themselves are mounted on 1.8-m-long standoffs, so putting all this together, the Tx antenna line needs to be 26.4m long, and the Rx antenna line needs to be 38.9m long. As they are both <60m in length, they can be made from 7/8-inch-diameter Heliax cable.

The remaining cables are all in the equipment rack that houses the base stations, the transmitter combiner, and the receiver multicoupler. To help determine the lengths of these cables, Figure 12.33 shows where the equipment is fitted in the rack. There is nothing complicated about this; generally, the size of the cavity filters in the transmitter combiner dictates that these go at the bottom of the rack. Then, to keep the transmit and receive filtering physically apart, the receiver filter and multicoupler would be fitted near the top of the rack along with the antenna system monitor. Lastly, the four base stations would go in the middle, between the transmit and receive filtering.

The rack used at this site is 2m high, so sensible lengths for the remaining cables would be:

- 1m, 0.8m, 0.6m, and 0.4m for the four transmitter jumpers;
- 1m, 0.8m, 0.6m, and 0.4m for the four receiver jumpers;

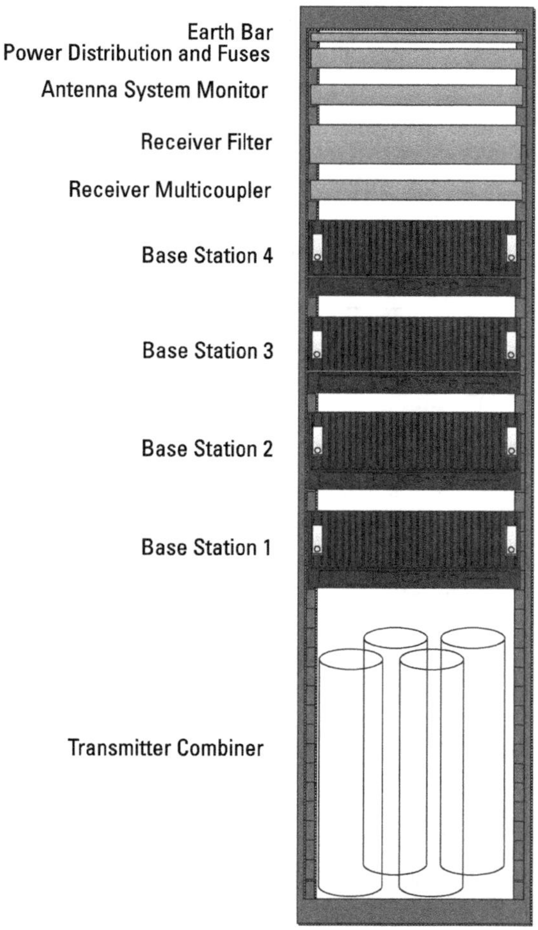

Figure 12.33 Hilltop Radio Station equipment rack layout.

- 1.6m for the transmitter combiner jumper;
- 0.4m for the receiver multicoupler jumper.

Those are the lengths of all the cables calculated now, so the transmission line specification for the Hilltop Radio Station site RF system would be as shown in Table 12.16.

12.7.1 RF Surge Suppression Devices

RF surge suppressors are designed to pass the RF signal at very low loss but protect the connected RF equipment by diverting any large surge (such as that caused by a lightning strike) to ground. To prevent the surge energy from entering the equipment shelter, the surge suppression devices are installed at the entry to the shelter at the junction between the Tx and Rx surge suppressor jumpers and their associated antenna lines as was shown in Chapter 11, Figure 11.3.

An example RF surge suppressor suitable for our system is the Polyphaser VHF50 (100–512 MHz). There would be two of them fitted at the shelter

Table 12.16
Hilltop Radio Station Transmission Line Specifications

Cable Name	Cable Type	Cable	Connector 1	Connector 2	Length/s
Tx Antenna Jumper Rx Antenna Jumper	1/2-inch Heliax	LDF4-50	7/16DIN (M)	7/16DIN (M)	3m 3m
Tx Antenna Line Rx Antenna Line	7/8-inch Heliax	AVA5-50	7/16DIN (F)	7/16DIN (F)	26.4m 38.9m
Tx Surge Suppressor Jumper	1/2-inch Heliax	LDF4-50	7/16DIN (M)	7/16DIN (M)	18m
Rx Surge Suppressor Jumper	1/2-inch Heliax	LDF4-50	Type N(M)	7/16DIN (M)	18m
Tx Combiner Jumper	1/2-inch Heliax	LDF4-50	7/16DIN (M)	7/16DIN (M)	1.6m
Rx Multicoupler Jumper	1/4-inch Superflex	FSJ1-50	Type N(M)	Type N(M)	0.4m
Transmitter Jumper	1/2-inch Superflex	FSJ4-50	Type N(M)	Type N(M)	1m, 0.8m, 0.6m, 0.4m
Receiver Jumper	1/4-inch Superflex	FSJ1-50	BNC (M)	BNC (M)	1m, 0.8m, 0.6m, 0.4m

entrance and grounded to the inside master ground bar: one in the transmit path between the Tx surge suppressor jumper and the Tx antenna line, and the other between the Rx antenna line and Rx surge suppressor jumper.

12.7.2 Transmission Line Grounding Devices

Transmission line grounding devices fit around to the outer conductor of the transmit and receive antenna lines and ground them to the tower leg. This ensures that the outer conductor remains grounded throughout the length of the antenna lines, minimizing insertion losses and preventing a large potential difference from developing on the transmission line in the event of a lightning strike.

For our Hilltop Radio Station system, according to the design rules given in Chapter 11, transmission line grounding devices would be installed at:

- The point of entrance to/exit from the equipment shelter;
- Near the bottom of the tower where the Tx or Rx antenna line makes the horizontal to vertical transition;
- The base of the transmit and receive antennas.

12.8 The Finished Design

We have designed the site RF system for Hilltop Radio Station. A full schematic drawing of the whole system would unfortunately be far too large to fit on one page of this book, so please see Figure 12.34, which at least shows where everything fits and references the detailed design drawings for each subsystem.

12.9 Concluding Remarks

Many thanks for your time reading this book. I hope it was helpful. I am sure that there is more than one way to carry out an interference-free repeater site RF system design for optimum coverage, but the methods presented in this book are the way that I have learned to do it over many years. It is knowledge that is even more relevant in today's crowded RF spectrum. In some way, I hope this book has inspired you to have a go; there is real satisfaction in getting it right and knowing that you have handed your customers a system that will not let them down at the time that they need it most.

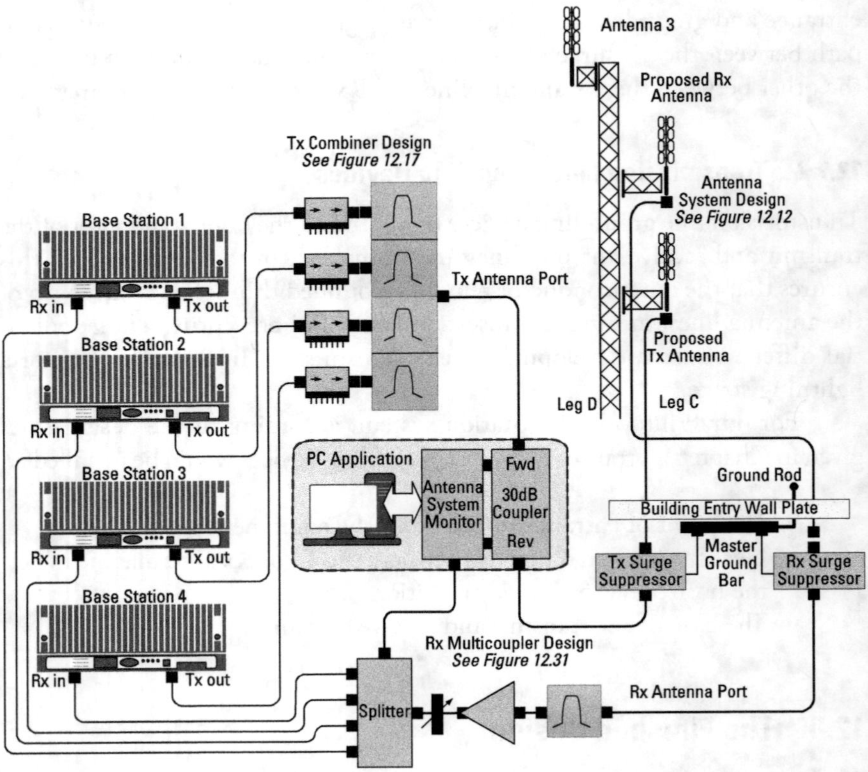

Figure 12.34 Hilltop Radio Station repeater site RF system design.

Selected Bibliography

International Telecommunications Union, Report ITU-R M.2244, Isolation Between Antennas of IMT Base Stations in the Land Mobile Service, November 2011.

RFI Technology Solutions, OA40-67 4 Element Offset Array Antenna, https://www.rfi.com.au/OA40-67-P.

RFI Technology Solutions, SPD4044-3115 Isolator/Circulator data sheet.

RFI Technology Solutions, 380-550MHz Cavity Filters data sheet, https://www.rfi.com.au/core/media/media.nl?id=1930303&c=3389566&h=vfSlLGN2tquCYDbkD9TzpeEf5umcgaMz46GNOTv0y3BYyBmS.

RFI Technology Solutions, 380-520MHz Receiver Multicoupler data sheet, https://www.rfi.com.au/core/media/media.nl?id=1930503&c=3389566&h=CE_f2kNckoJDG7gXr7LVUNfzn5X7qOvhVm43ScD2sQpf16gW.

RFI Technology Solutions, 400-520MHz UHF Preselector Filter data sheet, https://www.rfi.com.au/core/media/media.nl?id=3897414&c=3389566&h=AXCFTiyrQKADwP2myLnnM44CsOiwlyP7BfThoYJc8yE1qpOZ.

RFI Technology Solutions, Antenna System Monitor data sheet, https://www.rfi.com.au/core/media/media.nl?id=2490452&c=3389566&h=EKeZ39Wc7o6Dcgrxosqr_8BnkfspEBVj7-I3rub-7trVB5Ib.

Telecommunications Industry Association, Report TSB-88.1-F, Wireless Communications Systems—Performance in Noise and Interference Limited Situations—Part 1: Recommended Methods for Technology Independent Narrowband Performance Modelling, April 2022.

About the Author

Ian Graham is a principal engineer at Tait Communications in Christchurch, New Zealand. He has 40 years of circuit, product, and system-level RF design experience. He spent the formative years of his career at Marconi Avionics in the United Kingdom, learning and working on the RF transmitter and receiver designs for military projects before emigrating to New Zealand in 1994. Despite thinking the move was only for a couple of years, New Zealand quickly became home and Mr. Graham has now been at Tait Communications for 27 years. Initially, he worked on the RF design of base station, mobile and portable radio transmitters, and receivers in VHF, UHF, and 800-MHz bands and became an RF technology leader in 2003.

In a necessary shift in emphasis to system design, Mr. Graham moved into the role of Tait Global RF System Design SME in 2010. In that role, he has authored several Tait RF system design best-practice standards that have been fine-tuned ever since through leading the RF design of land mobile radio projects for public safety and utilities customers in the United States, Europe, Australia, and New Zealand.

Mr. Graham is passionate about passing on his knowledge and is responsible for training and mentoring Tait RF engineers around the world. It was never the intention to write a book, but the comments he received about this training formed the motivation and here we are.

About the Author

[illegible] New Zealand [illegible]

[illegible] leader in 2005.

[illegible]

[illegible] Australia and New Zealand.

[illegible]

Index

Artech House Mobile Communications Library

William Webb, Series Editor

3G CDMA2000 Wireless System Engineering, Samuel C. Yang

3G Multimedia Network Services, Accounting, and User Profiles, Freddy Ghys, Marcel Mampaey, Michel Smouts, and Arto Vaaraniemi

5G and Satellite RF and Optical Integration, Geoff Varrall

5G-Enabled Industrial IoT Networks, Amitava Ghosh, Rapeepat Ratasuk, Simone Redana, and Peter Rost

5G New Radio: Beyond Mobile Broadband, Amitav Mukherjee

5G Spectrum and Standards, Geoff Varrall

802.11 WLANs and IP Networking: Security, QoS, and Mobility, Anand R. Prasad and Neeli R. Prasad

Achieving Interoperability in Critical IT and Communications Systems, Robert I. Desourdis, Peter J. Rosamilia, Christopher P. Jacobson, James E. Sinclair, and James R. McClure

Advances in 3G Enhanced Technologies for Wireless Communications, Jiangzhou Wang and Tung-Sang Ng, editors

Advances in Mobile Information Systems, John Walker, editor

Advances in Mobile Radio Access Networks, Y. Jay Guo

Artificial Intelligence in Wireless Communications, Thomas W. Rondeau and Charles W. Bostian

Broadband Wireless Access and Local Network: Mobile WiMax and WiFi, Byeong Gi Lee and Sunghyun Choi

CDMA for Wireless Personal Communications, Ramjee Prasad

CDMA RF System Engineering, Samuel C. Yang

CDMA Systems Capacity Engineering, Kiseon Kim and Insoo Koo

Cell Planning for Wireless Communications, Manuel F. Cátedra and Jesús Pérez-Arriaga

Cellular Communications: Worldwide Market Development, Garry A. Garrard

Cellular Mobile Systems Engineering, Saleh Faruque

Cognitive Radio Interoperability through Waveform Reconfiguration, Leszek Lechowicz and Mieczyslaw M. Kokar

Cognitive Radio Techniques: Spectrum Sensing, Interference Mitigation, and Localization, Kandeepan Sithamparanathan and Andrea Giorgetti

The Complete Wireless Communications Professional: A Guide for Engineers and Managers, William Webb

Designing RF Combining Systems for Shared Radio Sites, Ian Graham

EDGE for Mobile Internet, Emmanuel Seurre, Patrick Savelli, and Pierre-Jean Pietri

Emerging Public Safety Wireless Communication Systems, Robert I. Desourdis, Jr., et al.

From LTE to LTE-Advanced Pro and 5G, Moe Rahnema and Marcin Dryjanski

The Future of Wireless Communications, William Webb

Geospatial Computing in Mobile Devices, Ruizhi Chen and Robert Guinness

GPRS for Mobile Internet, Emmanuel Seurre, Patrick Savelli, and Pierre-Jean Pietri

GSM and Personal Communications Handbook, Siegmund M. Redl, Matthias K. Weber, and Malcolm W. Oliphant

GSM Networks: Protocols, Terminology, and Implementation, Gunnar Heine

GSM System Engineering, Asha Mehrotra

Handbook of Land-Mobile Radio System Coverage, Garry C. Hess

Handbook of Mobile Radio Networks, Sami Tabbane

Handbook of Next-Generation Emergency Services, Barbara Kemp and Bart Lovett

High-Speed Wireless ATM and LANs, Benny Bing

Implementing Full Duplexing for 5G, David B. Cruickshank

In-Band Full-Duplex Wireless Systems Handbook, Kenneth E. Kolodziej, editor

Mobile IP Technology for M-Business, Mark Norris

Mobile Satellite Communications, Shingo Ohmori, Hiromitsu Wakana, and Seiichiro Kawase

Mobile Telecommunications Standards: GSM, UMTS, TETRA, and ERMES, Rudi Bekkers

Mobile-to-Mobile Wireless Channels, Alenka Zajić

Mobile Telecommunications: Standards, Regulation, and Applications, Rudi Bekkers and Jan Smits

Multiantenna Digital Radio Transmission, Massimiliano Martone

Multiantenna Wireless Communications Systems, Sergio Barbarossa

Multi-Gigabit Microwave and Millimeter-Wave Wireless Communications, Jonathan Wells

Multiuser Detection in CDMA Mobile Terminals, Piero Castoldi

OFDMA for Broadband Wireless Access, Slawomir Pietrzyk

Practical Wireless Data Modem Design, Jonathon Y. C. Cheah

The Practitioner's Guide to Cellular IoT, Cameron Kelly Coursey

Prime Codes with Applications to CDMA Optical and Wireless Networks, Guu-Chang Yang and Wing C. Kwong

Quantitative Analysis of Cognitive Radio and Network Performance, Preston Marshall

QoS in Integrated 3G Networks, Robert Lloyd-Evans

Radio Resource Management for Wireless Networks, Jens Zander and Seong-Lyun Kim

Radiowave Propagation and Antennas for Personal Communications, Third Edition, Kazimierz Siwiak and Yasaman Bahreini

RDS: The Radio Data System, Dietmar Kopitz and Bev Marks

Resource Allocation in Hierarchical Cellular Systems, Lauro Ortigoza-Guerrero and A. Hamid Aghvami

RF and Baseband Techniques for Software-Defined Radio, Peter B. Kenington

RF and Microwave Circuit Design for Wireless Communications, Lawrence E. Larson, editor

Sample Rate Conversion in Software Configurable Radios, Tim Hentschel

Signal Processing Applications in CDMA Communications, Hui Liu

Signal Processing for RF Circuit Impairment Mitigation, Xinping Huang, Zhiwen Zhu, and Henry Leung

Smart Antenna Engineering, Ahmed El Zooghby

Software-Defined Radio for Engineers, Travis F. Collins, Robin Getz, Di Pu, and Alexander M. Wyglinski

Software Defined Radio for 3G, Paul Burns

Software Defined Radio: Theory and Practice, John M. Reyland

Spectrum Wars: The Rise of 5G and Beyond, Jennifer A. Manner

Spread Spectrum CDMA Systems for Wireless Communications, Savo G. Glisic and Branka Vucetic

Technical Foundations of the IoT, Boris Adryan, Dominik Obermaier, and Paul Fremantle

Technologies and Systems for Access and Transport Networks, Jan A. Audestad

Third-Generation and Wideband HF Radio Communications, Eric E. Johnson, Eric Koski, William N. Furman, Mark Jorgenson, and John Nieto

Third Generation Wireless Systems, Volume 1: Post-Shannon Signal Architectures, George M. Calhoun

Traffic Analysis and Design of Wireless IP Networks, Toni Janevski

Transmission Systems Design Handbook for Wireless Networks, Harvey Lehpamer

UMTS and Mobile Computing, Alexander Joseph Huber and Josef Franz Huber

Understanding Cellular Radio, William Webb

Understanding Digital PCS: The TDMA Standard, Cameron Kelly Coursey

Understanding WAP: Wireless Applications, Devices, and Services, Marcel van der Heijden and Marcus Taylor, editors

Universal Wireless Personal Communications, Ramjee Prasad

Virtualizing 5G and Beyond 5G Mobile Networks, Larry J. Horner, Kurt Tutschku, Andrea Fumagalli, and ShunmugaPriya Ramanathan

WCDMA: Towards IP Mobility and Mobile Internet, Tero Ojanperä and Ramjee Prasad, editors

Wi-Fi 6: Protocol and Network, Susinder R. Gulasekaran and Sundar G. Sankaran

Wireless Communications in Developing Countries: Cellular and Satellite Systems, Rachael E. Schwartz

Wireless Communications Evolution to 3G and Beyond, Saad Z. Asif

Wireless Intelligent Networking, Gerry Christensen, Paul G. Florack, and Robert Duncan

Wireless LAN Standards and Applications, Asunción Santamaría and Francisco J. López-Hernández, editors

Wireless Sensor and Ad Hoc Networks Under Diversified Network Scenarios, Subir Kumar Sarkar

Wireless Technician's Handbook, Second Edition, Andrew Miceli

For further information on these and other Artech House titles, including previously considered out-of-print books now available through our In-Print-Forever® (IPF®) program, contact:

Artech House
685 Canton Street
Norwood, MA 02062
Phone: 781-769-9750
Fax: 781-769-6334
e-mail: artech@artechhouse.com

Artech House
16 Sussex Street
London SW1V 4RW UK
Phone: +44 (0)20 7596-8750
Fax: +44 (0)20 7630-0166
e-mail: artech-uk@artechhouse.com

Find us on the World Wide Web at: www.artechhouse.com